Routledge Revivals

The International Atomic Energy Agency and World Nuclear Order

The International Atomic Energy Agency has had a leading responsibility in preventing the spread of nuclear weapons and misuse of materials intended for nuclear energy across the world. Originally published in 1987, in the wake of the Chernobyl disaster of 1986 which proved the utmost importance of the agency, Scheinman explores the function of the IAEA and the challenges it faced. This report also lists ways that the agency could be strengthened touching on topics such as leadership roles, support for safeguarding functions and prevention from the agency being overwhelmed by international and national political issues. This title will be of interest to students of Environmental Studies.

The International Atomic Energy Agency and World Nuclear Order

Lawrence Scheinman

Routledge
Taylor & Francis Group

First published in 1987
by Resources for the Future, Inc.

This edition first published in 2016 by Routledge
2 Park Square, Milton Park, Abingdon, Oxon, OX14 4RN
and by Routledge
711 Third Avenue, New York, NY 10017

Routledge is an imprint of the Taylor & Francis Group, an informa business

A Library of Congress record exists under LC control number: 87042832

ISBN 13: 978-1-138-95228-7 (hbk)
ISBN 13: 978-1-315-66767-6 (ebk)
ISBN 13: 978-1-138-95232-4 (pbk)

THE INTERNATIONAL ATOMIC ENERGY AGENCY *and* WORLD NUCLEAR ORDER

THE INTERNATIONAL ATOMIC ENERGY AGENCY *and* WORLD NUCLEAR ORDER

LAWRENCE SCHEINMAN

Resources for the Future • *Washington, DC*

Printed in the United States of America

Published by Resources for the Future, Inc.
1616 P Street, N.W., Washington, D.C. 20036

Books from Resources for the Future are distributed worldwide by The Johns Hopkins University Press

Library of Congress Cataloging-in-Publication Data

Scheinman, Lawrence.
The International Atomic Energy Agency and world nuclear order.

Includes index.
1. International Atomic Energy Agency. 2. Nuclear energy—Government policy—United States. 3. Nuclear nonproliferation. I. Title.
QC770.I4962S34 1987 333.79'26'0601 87-42832
ISBN O-915707-35-7 (alk. paper)
ISBN O-915707-36-5

♾ The paper in this book meets the guidelines for permanence and durability of the Committee on Production Guidelines for Book Longevity of the Council on Library Resources.

RESOURCES FOR THE FUTURE (RFF) is an independent nonprofit organization that advances research and public education in the development, conservation, and use of natural resources and in the quality of the environment. Established in 1952 with the cooperation of the Ford Foundation, it is supported by an endowment and by grants from foundations, government agencies, and corporations. Grants are accepted on the condition that RFF is solely responsible for the conduct of its research and the dissemination of its work to the public. The organization does not perform proprietary research.

RFF research is primarily social scientific, especially economic, and is concerned with the relationship of people to the natural environment—the basic resources of land, water, and air; the products and services derived from them; and the effects of production and consumption on environmental quality and human health and well-being. Grouped into three research divisions—Energy and Materials, Quality of the Environment, and Renewable Resources—staff members pursue a wide variety of interests, including food and agricultural policy, forest economics, natural gas policy, multiple use of public lands, mineral economics, air and water pollution, energy and national security, hazardous wastes, and the economics of outer space. Resident staff members conduct most of the organization's work; a few others carry out research elsewhere under grants from RFF.

Resources for the Future takes responsibility for the selection of subjects for study and for the appointment of fellows, as well as for their freedom of inquiry. The views of RFF staff members and the interpretations and conclusions of RFF publications should not be attributed to Resources for the Future, its directors, or its officers. As an organization, RFF does not take positions on laws, policies, or events, nor does it lobby.

Lawrence Scheinman is Professor of Government at Cornell University.

This book was designed by Martha Ann Bari. The index was prepared by Baehr Publishing Services.

CONTENTS

TABLES

FIGURES

FOREWORD

This book is a major event in a multifaceted program of meetings, conferences, seminars, and briefings to increase understanding of the importance of the International Atomic Energy Agency (IAEA) to the foreign policy of the United States and to the maintenance of the peace and security of the world.

In 1984 when the program was initiated, the IAEA was in trouble. Even the United States, which was the parent of the IAEA, was no longer providing the unequivocal support and leadership that it had in the past. Evidence of IAEA's troubles included:

- Congress was cutting the administration's budget requests for the IAEA.
- The IAEA had become increasingly an arena for political contentions extraneous to the international management of nuclear affairs.
- Many IAEA members were becoming restless about how the agency was operating and the balance between its two principal activities—the technical assistance and safeguards programs.
- The agency faced increased challenges in Congress and elsewhere to the effectiveness of IAEA safeguards and diminished confidence in them and in the agency itself.

These matters were of serious concern to those who participated in this program because they believe that the safeguards program of the IAEA is a linchpin in the nonproliferation regime and that the possible spread of nuclear weapons to additional countries poses grave threats to the security of the United States and to the peace of the world.

The United States has a large stake in the effectiveness of the IAEA and in its capacity to develop capabilities commensurate with the growing concern about nuclear safety, the spread of nuclear weapons, and the continuing growth of nuclear energy.

Because the United States initiated the agency, is the largest contributor (25 percent), has been the principal shaper of agency programs and practices throughout its history, and remains its most influential member, the departure from its historic role of positive leadership is all the more telling on the quality and performance of the agency. It means that the questions that have arisen about the credibility of agency safeguards, the politicization of the membership, the ability to maintain and improve its performance in a steadily increasing number of nuclear facilities, and similar problems will continue to remain unresolved. The possibility must be faced that, for the first time, the quality of the IAEA's performance may begin to decline, much to the detriment of U.S. and global interests in world peace and security.

The tragic accident at Chernobyl in the Ukraine is evidence of the singular importance of the IAEA in international nuclear affairs. It was the forum for international assessment of the causes and consequences of the accident and for the enactment of two new international conventions on nuclear safety and accident management. In the same vein, let us hope that it does not take a further proliferation of nuclear weapons to remind the world of what might have been had it been more attentive to the IAEA.

I wish to express my appreciation to Resources for the Future for providing a home for this program and giving professional support for its activities; to the distinguished members of the Project Advisory Group (listed below) for their participation in many of our educational activities and for their overall guidance and advice; to our author, Lawrence Scheinman, who devoted considerable time and effort to the research and analysis of the materials and interviews on which this book is based and for the preparation of a monograph ("The Nonproliferation Role of the International Atomic Energy Agency: A Critical Assessment") that proved invaluable in our educational ventures; and to the many others whose help has been so valuable. Throughout this project my friend and associate, Herman Pollack, has served as project manager with devotion and initiative. Finally, I must express my deep appreciation to those several friends from my Cleveland days who willingly provided their financial support for the project, and, as well, to the Rockefeller Brothers Fund, the W. Alton Jones Foundation, and the Carnegie Corporation for their generous support.

This RFF book highlights the origins and development of the IAEA, the issues now confronting it, and its critical role in the international nonproliferation regime. It contains, as well, a number of suggestions that, if pursued, we believe will pay immense dividends in the continuing quest for world peace and tranquility.

Washington, D.C. *T. Keith Glennan*
July 1987

PROJECT ADVISORY GROUP

T. KEITH GLENNAN, Chairman, former U.S. Representative to the IAEA

RUTH ADAMS, former Editor, *Bulletin of the Atomic Scientists*

WARREN DONNELLY, Senior Research Associate, Congressional Research Service

THEODORE HESBURGH, S.J., President Emeritus, Notre Dame University

ASHTON J. O'DONNELL, Vice President, Bechtel Corporation

HERMAN POLLACK, Research Professor of International Affairs, George Washington University

JAMES R. SCHLESINGER, former Secretary of Energy, Secretary of Defense, Director of the CIA, and Chairman of the Atomic Energy Commission

GLENN T. SEABORG, Nobel laureate in chemistry

GERALD C. SMITH, President of the Arms Control Association and former U.S. Representative to the IAEA

HENRY DEWOLF SMYTH, former U.S. Representative to the IAEA

GERALD F. TAPE, former U.S. Representative to the IAEA

ACKNOWLEDGMENTS

Although many debts have been incurred in writing this book, it could not have been undertaken without the inspiring and unflagging leadership of T. Keith Glennan, who initiated the project, brought together a distinguished panel of advisors, and marshaled the necessary resources to research and write this study. Many people contributed in one way or another to the final product, and while there are too many to name individually, several deserve special mention for their efforts. My thanks to Herman Pollack for his enormous wisdom, compassion, and encouragement, all of which were important to me. Warren Donnelly offered invaluable advice and recommendations on substance and style. Gerald Tape, Ashton O'Donnell, and Gerard Smith all played important advisory roles.

Special thanks are due to Joseph Pilat who provided an initial draft of chapter 2 and commented extensively on several other parts of the text, and to Abraham Friedmann who prepared an initial draft of the third chapter and did yeoman service in commenting on other chapters. My safeguards chapters benefited substantially from generous comments offered by James de Montmollin and Eugene Weinstock whose sage advice was not always taken. Mea culpa. Important contributions and correctives were made by Leonard Weiss, Leonard Spector, Charles van Doren, David Fischer, and Alan Labowitz, and by the anonymous readers who juried the manuscript for Resources for the Future. Particular thanks are due to John Ahearne of RFF, a good friend and indomitable and constructive critic. I am also indebted to my son Adam who scrutinized much of the relevant literature to ensure my awareness of all that had been written on the IAEA.

The study is based not only on documents and literature, but on more than fifty interviews conducted over a two-and-a-half year period commencing in late 1984. I owe much to the patient and helpful role played by many IAEA staff members and members of permanent missions to the IAEA. Their collective experience and wisdom enriched my understanding of IAEA activities and problems even beyond what I had learned in the mid and late 1970s as an official of the Department of State and of the Energy Research and Development Administration. But in the final analysis, I must return to Keith Glennan, whose involvement has meant so much to me and to the book. I can only hope that the resulting product will contribute to greater understanding of the IAEA and its importance to international nuclear cooperation, to nonproliferation, and to achieving the hopes for a better and more secure world that lie at the core of the agency's statute and of Keith Glennan's concerns. In a very personal sense, it had better do so, considering the tolerance that my wife Lorraine (to whom this book is dedicated) showed for the many absences from home and the run of deferred holidays that this book entailed. If it helps to advance the public policy dialogue on international cooperation and control of the peaceful atom, it will have been worth the price to both of us.

May 1987 *Lawrence Scheinman*

THE INTERNATIONAL ATOMIC ENERGY AGENCY *and* WORLD NUCLEAR ORDER

Chapter 1

INTRODUCTION AND OVERVIEW

The great challenge to humanity is how to avoid nuclear war. Understandably, mention of this threat invariably focuses our attention on the political competition and nuclear weapons arsenals of the superpowers and on their progress, or lack thereof, in bringing the nuclear arms race under control. The risk of nuclear war is primarily a question of the political and nuclear weapons policies and postures of the United States and the Soviet Union. Unilateral force management, bilateral relations (including arms control agreements and crisis management capabilities), and attitudes of political leaders in these two nations toward the usability of nuclear weapons and the plausibility of controlled or limited nuclear war are the predominant factors here.

The improbability of nuclear war does not rule out its possibility. If this is true with respect to the superpowers—with their awesome and sophisticated arsenals, command and control systems, and retaliatory capabilities—it is even more likely true of lesser nuclear nations. Proliferation can have unsettling effects on local, regional, and global stability and security, and it will complicate the delicate nuclear balance that exists today. Although proliferation need not have that effect in every instance, it may, and many policymakers have been inclined to think that it likely would. This is principally why so much effort has been devoted for so long to seeking to prevent the proliferation of nuclear weapons to those countries that do not now have them. This is not to argue that there is an automatic linkage or direct causality between proliferation and nuclear war. That relationship is a function of many factors, including who is doing the proliferation, the circumstances under which it takes place, and the impact on the behavior and political decisions of other states. But to be fully appreciated, the whole issue of proliferation must be seen against the broader background of the problem of nuclear war.

Although nuclear proliferation has been the object of policy attention in a number of countries, and particularly the United States, it is not as prominent as Star Wars or the U.S.-Soviet arms race, and the attention it receives is not always commensurate with the dangers it can pose to national and international security and stability. Nonproliferation has tended to exist as a netherworld, sometimes part of security and foreign policy, sometimes not. If this is true for the United States, where proliferation has been of concern at the highest levels of government, and has been on the political agenda since the dawn of the nuclear age, it is even more true for other countries. Raising the consciousness of the political community and attentive public is one of the primary objectives of this book.

Another, more immediate, and more central purpose is to explore and evaluate the nonproliferation role of the International Atomic Energy Agency (IAEA). For more than a quarter-century, the IAEA has had a leading responsibility in national and international efforts to stem the further spread of nuclear weapons and to deter misuse of facilities and materials intended for civil nuclear purposes. The principal nonproliferation function of the IAEA is to verify by safeguards—including on-site inspection—that safeguarded facilities or materials or both are being used only for peaceful purposes and not being diverted to make nuclear weapons. This unique and vital function places the IAEA at the heart of the nonproliferation regime that is intended to avoid the further spread of nuclear weapons.

Today, the IAEA faces problems that threaten to discredit the effectiveness and reliability of its safeguards and subject it to the loss of confidence plaguing other international bodies, as exemplified by the U.S. withdrawal from UNESCO in 1985.[1] Safeguards alone are not determinants of national proliferation. That, as discussed below, is a matter of hard political decision based on many factors. However, the discrediting or ultimate failure of IAEA safeguards could severely weaken the nonproliferation regime. The demise of the regime could confront the international community with a new surge of nuclear weapons by nations whose political and security calculus could no longer count on the objective international verification, now provided by the IAEA safeguards system, that civil nuclear technology was not being diverted or abused.[2] The difficulties confronting the IAEA risk endangering the entire nonproliferation regime, national security, and international stability. The overall goal of this analysis is to provide a background for public appreciation of the agency and its future, and to stimulate hard thinking on how the unique qualities and features of the IAEA can be most effectively exploited.

Our analysis is addressed to the many who, although not part of the nuclear community, hold the future of the IAEA in their hands. Without their recognition and support for its purposes, the fate of the agency could be overshadowed by the more dramatic issues now on the international political stage. Here, we will explore the trends and developments affecting the agency, and examine its principal strengths and weaknesses as well as its limitations and potential for assuring the peaceful use of nuclear materials as a source of energy. We also will analyze the implications of the trends and problems for the future efficacy of the IAEA and its international safeguards system. Finally, we will offer some recommendations for strengthening confidence in the IAEA and its safeguards system, and for enhancing the institution's credibility.[3]

PLAN OF THE BOOK

Chapter 2 discusses the background and origins of the IAEA from the Baruch Plan, through Atoms for Peace, to negotiations resulting in the establishment of the agency. Chapter 3 provides an overview of the structure and programmatic activities of the IAEA. Emphasis is given to special structural characteristics of the agency and their evolution, in particular the board of governors, the general conference, and the secretariat. Programmatic discussion is focused primarily on the facilitative side, in particular, technical cooperation and assistance. The budget of the agency also is discussed here.

Chapters 4 and 5 deal with the agency's safeguards system. Chapter 4 discusses the development of safeguards from the inception of the agency to the time it assumed safeguards responsibilities under the 1970 Treaty on the Non-Proliferation of Nuclear Weapons (NPT), with particular attention to the non-NPT safeguards document (INFCIRC/66) that still applies today to a number of potential proliferator states including Pakistan, India, Argentina, Brazil, South Africa, Israel, and Spain. Chapter 5 focuses primarily on the NPT safeguards document (INFCIRC/153), examining the differences and similarities with the earlier document (INFCIRC/66) and the impact of a broadened responsibility on the agency. Together, the chapters look into the objectives of safeguards, implementation experience, and the problems that have arisen.

Chapter 6 reviews the key events of the mid and late 1970s (for example, India's nuclear test, the imminent spread of sensitive nuclear technologies to unstable regions of the world, the growing competitiveness of the world nuclear market) that contributed to a change in national nonproliferation policies in certain key states, particularly the United States, and considers the impact of those policy changes on the IAEA, international safeguards, and the nonproliferation regime.

Chapter 7 examines four key problems that have confronted the IAEA as a consequence of the changing international environment described in the previous chapter, and considers the effect that they have had on the agency's ability to fulfill its statutory responsibilities and the impact of that on the international nuclear regime.

Chapter 8, which is the concluding chapter, summarizes the key aspects of the preceding analysis and offers a prospectus on the future. It proposes measures or actions in four areas—leadership, safeguards, depoliticization, and sustaining balance between the agency's two central responsibilities—in the interest of ensuring successful continuation of the IAEA as a central cog in the machinery of international nuclear cooperation and nonproliferation.

SOME THOUGHTS ON NONPROLIFERATION

Several considerations are important background for understanding the IAEA's nuclear responsibilities. These include: (1) examination of why states decide to forgo or retain the option to acquire nuclear weapons; (2) insight into the potential linkage between civil nuclear power and nuclear weapons; and (3) awareness of the nature and scope of the nonproliferation regime itself. Accordingly, the balance of this chapter is devoted largely to these questions.[4]

Motivations to Retain the Nuclear Option

Although some would argue that the spread of nuclear weapons could produce the same deterrent effect between other rival countries that has emerged between the superpowers,[5] the prevalent view has been that, on balance, further proliferation likely would have profoundly disruptive effects on world peace and order. Proliferation, in this view, would add new and dangerous dimensions to existing disputes between nations. It could invite preemptive attacks, or at least provoke regional nuclear arms races that not only would increase the prospects that nuclear weapons might be used in regional conflicts, but, to the extent that the superpowers became involved in such situations, would raise the possibility of cataclysmic nuclear war.[6] Furthermore, proliferation would complicate efforts to negotiate acceptable and effective arms control arrangements, and generally work to the disadvantage of international stability. Avoiding the spread of nuclear weapons, then, is an essential concomitant to U.S.-Soviet efforts to achieve agreement on the vital question of nuclear arms control and disarmament. The fortunate convergence of superpower interests here also contributes to stabilizing relations between them.

One hundred twenty-six non-nuclear weapon states have ratified the 1970 treaty and thereby formally pledged to forgo the acquisition or development of nuclear weapons and to accept international inspection of their nuclear activities by the IAEA.[7] These states apparently share the view that their security interests are better served by abstaining from nuclear weapons, at least under present circumstances, than by acquiring them. Not all states, unfortunately, have elected to make legally binding multilateral nonproliferation commitments. Regrettably, among those who have not are a number of states whose advanced nuclear capabilities raise the possibility of their becoming nuclear weapon states if they make that political choice. Among these are Israel, South Africa, Pakistan, India, and Argentina. Even among states in this category, however, none are acknowledged nuclear weapon states, and all deny any intent to acquire such weapons.

In most cases, the motive for retaining the nuclear option springs from political or military security threats and perceived compensatory advantages from having nuclear weapons or the capability to produce them quickly without abrogating contrary legal obligations. Israel and South Africa, nations that are largely politically isolated and confronted with hostile neighbors seeking to alter the status quo, view maintenance of the nuclear option as the best way to preserve their security or to maintain leverage in eventually negotiating a politically acceptable *modus vivendi*. Except to state that they would not be the first to introduce nuclear weapons into their area, neither sees any reason to formally commit to nonproliferation except in the context of a general political settlement that removes existing threats to their security and results in regional stability.[8]

Pakistan, confronted by a much larger neighbor with an already proven nuclear explosive capability, and with whom it has fought and lost several bitter wars since the two nations gained their independence in 1947, also sees nuclear capability and preservation of the nuclear option as essential to its current security situation and to establishing a deterrent to potential future aggression.[9] For its part, India, the larger neighbor, has had fewer security reasons to eschew signing the NPT. This is so despite the losses sustained in the 1962 Sino-Indian war and subsequent entry of the People's Republic of China into the nuclear weapons club in 1964. Internal pressures to proliferate were stimulated by those events but were rejected. Up to the present, India has confined itself to demonstrating its ability to construct and test a nuclear explosive device. However, it has insisted that the 1974 test was a peaceful nuclear explosion, carried out consistent with the country's international commitments (including its obligation under the 1963 Limited Test Ban Treaty), and not an act of weapons proliferation. Nevertheless, India is accumulating weapons-usable material that for reasons of its purely indigenous origin is subject neither to restrictive use limitations nor to IAEA safeguards.[10]

Concern with Pakistan's emerging nuclear capabilities (earlier disparaged by Indian analysts) may alter the Indian position and make security a still more significant consideration in the Indian nuclear policy equation. But India, like Argentina (which faces even fewer direct security threats), has a broader agenda. Both nations aspire to positions of regional leadership and world importance, positions they do not believe can be secured if they acquiesce to conditions or restrictions that are not universally applicable.[11] To them, the NPT is a discriminatory instrument that confirms and legitimizes two classes of nations. One class, the nuclear weapon nations, is placed in a position of permanent superiority, admonished but not obligated to limit and eventually terminate nuclear arms and subject only to such safeguards as those to which the governments voluntarily submit. All the rest are left in a position of permanent inferiority, foreclosed from developing peaceful nuclear explosive programs and obliged to submit all of their nuclear activities to international safeguards.[12] In the view of India and Argentina, such an agreement could not be accepted as a matter of political principle; nor (in their view) does refusal to become a party to such a treaty mean that its basic principles will be defied. To the contrary, they contend that one can be committed to nonproliferation principles and policy without endorsing discriminatory arrangements.[13]

Even this brief statement about motivation to maintain (or at least not foreclose) a nuclear weapons option cannot fail to highlight the importance to the success of nonproliferation of removing as many incentives to proliferate as possible. Political incentives are more difficult to cope with than are security incentives because they are less tangible and are tied to considerations of prestige, status, and international standing, which are largely matters of perception. As long as the nuclear weapon states pursue policies that are seen to confirm the value of maintaining (not to speak of expanding) nuclear arsenals, or, worse, policies that suggest, through their rhetoric or force structure, that nuclear weapons can be "conventionalized" and used to fight limited wars, it seems inevitable that governments that have sought to retain the nuclear weapons option for security reasons will be confirmed in their original policy judgment. And nations with political agendas involving regional leadership or predominance, or committed to repudiating rather than sustaining discriminatory arrangements that preserve privileges for the few, will be encouraged to continue to resist formal renunciation of the right to acquire nuclear weapons. Moreover, nations that have accepted the nuclear bargain of the NPT may lose patience with the failure of the superpowers to make any real progress in nuclear arms control, and either defect or refuse to vote in favor of continuing the NPT when in 1995 the parties meet to decide whether to keep the NPT in force. Thus, from the political perspective, implementation of commitments concerning cessation of the nuclear arms race, including agreement on a comprehensive test ban by the nuclear weapon nations party to the NPT, is something that cannot be ignored except at peril.[14]

From the security perspective, the challenge is how to remove the security incentive to proliferate, especially with regard to the nonaligned nations. Paradoxically, it is the existence of nuclear-anchored security guarantees embedded in alliance arrangements that accounts for the decision of many nations to forswear acquiring nuclear weapons, even though they have the technical capability. Individual or collective extension by the superpowers of formal, binding, and credible security guarantees to countries subject to significant security risks is one approach, but this is more often than not complicated by other political and security considerations. Devising acceptable and credible arrangements for nuclear-weapon-free zones, preferably, but not necessarily, coupled with adherence to instruments involving formal renunciation of nuclear weapons, is another. These are not new ideas, and their difficulty is underscored by the lack of significant progress on them over the past years. The hard truth is that there are no simple or obvious answers and no generic approaches to the problem. Each potential proliferator presents a separate problem and challenge for which a relevant and practical approach must be tailored. The way is marked by uncertainties.

While the success or failure of nonproliferation policies does not stand or fall on the decision to acquire nuclear weapons by one or two nonsignatories to the NPT, the strength of the no-weapons pledge increases each time an additional country chooses to make it. Each new nonproliferation pledge reinforces others and complicates decisions of the remaining nonparticipants to cut against the grain of a widely shared international ethic. Widespread international support for nonproliferation also raises the political costs of defecting from commitments anchored in treaties such as NPT or Tlatelolco (which established a nuclear-weapon-free zone in Latin America). On balance, during the past four nuclear decades, the international community has come to see more clearly the importance of a broadly based endorsement for nonproliferation. But this substantial consensus is continually subject to erosion and can be maintained only at the cost of eternal vigilance and unremitting commitment.

Linking Nuclear Power, Nuclear Weapons, and Nonproliferation

If the key to nonproliferation were simply to foreclose access to a particular kind of military weapon whose characteristics were unique and whose production could be easily isolated, the problem would be more straightforward. This, however, is not the case. Avoiding nuclear proliferation is complicated by two closely related factors.

First, the commercial energy argument: Nuclear science and technology not only gave humans control over awesome destructive power, it also opened the promise of a wholly new source of energy to support economic growth and development and social modernization, particularly for those nations lacking conventional energy resources. Nuclear energy is, therefore, of great interest to many countries. The appeal of nuclear energy is firmly established and not easily disrupted, despite the fact that its early promise may have been overstated initially. Its appeal was further undermined as problems of managing nuclear energy became more apparent, and as the number of countries with the industrial infrastructure necessary to take advantage of nuclear energy proved to be limited.

Second, the technology transfer argument: Up to a point the peaceful and military uses of nuclear energy share some of the same basic technology, materials, and infrastructure; non-weapon nations with active nuclear power programs, given the political motivation, could adapt them to military purposes. Even if nuclear materials and facilities were not diverted from civil uses, the accumulated knowledge and hands-on experience with nuclear technologies could aid military missions. Furthermore, reactor-grade plutonium can be made into nuclear explosives by non-nuclear weapon countries, although with a loss of yield and reliability. Thus the control of plutonium from reprocessing plants and in spent fuel storage is also a concern.

The first of these two factors is largely self-explanatory. The promise of potential economic and energy security benefits from nuclear energy has been of compelling importance to those interested in reducing dependence on imported conventional fuels. This long-standing interest was reinforced by the oil crisis of 1973–1974, which vividly demonstrated the effects of vulnerability to price manipulation and politically inspired embargoes. Even where current exploitation of nuclear power has not proved feasible, many countries seek to ensure freedom to fully exploit the nuclear alternative (including research, development, and demonstration as well as production) and consequently resist restraints that may limit or curtail that opportunity. For most of these countries, with no interest in nuclear weapons, the fact that nuclear power development may involve access to technologies and materials that have potential military value—such as reprocessing and plutonium—is not disturbing. In their view, the matter is resolved by their agreeing to nonproliferation undertakings and to international safeguards to verify their commitments, which is all that the NPT requires. Many of them vigorously resist suggestions, coming mainly from U.S. analysts (who are seen as reflecting the views of a nation richly endowed in the entire spectrum of energy resources), to refrain from exploring or using the full range of nuclear technologies and materials, including those that are defined as sensitive. They do not find research, development, demonstration, and eventual use of such materials and technologies to be inconsistent with support for nonproliferation.

Energy-short advanced industrial states (such as Japan and the Federal Republic of Germany) have proved to be particularly obdurate on this matter. And for their part, developing countries choose to view attempts at limiting access to sensitive nuclear technologies in anti-colonialist terms. They are disposed to interpret supplier-state restraint in transferring sensitive or high-risk technologies as neo-imperialism in which technology, rather than trade or investment, provides the leverage. Some of the loudest protests come from developing countries (such as India and Argentina) that, having rejected the NPT as discriminatory, also refuse to voluntarily place all of their nuclear activities under IAEA safeguards.

They nevertheless can count on the support of other Third World NPT states, which regard them as articulate advocates of the principle of full access to all nuclear technologies developed and exploited by the advanced industrial states, something the Third World at large views as key to advanced status. That many of these countries lack the industrial base to make effective use of these sophisticated nuclear technologies is irrelevant to their demands for open access.[15] Such considerations complicate efforts to achieve global consensus on constraints in nuclear fuel cycle development in support of nonproliferation more than most advocates of constraint have recognized.

The second (technological) factor raises directly the question of linkage between nuclear power and nuclear weapons. Sharply different views obtain on the extent to which such linkage exists. On the one side can be found arguments to the effect that once embarked on a nuclear power program a nation also has taken the first steps toward the bomb; that civil nuclear power programs can be convenient legitimizing covers for nuclear weapons programs; or that the existence of nuclear capabilities intended for peaceful purposes can be appropriated under pressure of changing political conditions to weapons uses.[16]

On the other side can be found the view that (1) while the nuclear power fuel cycle can provide the ingredients for nuclear weapons, what it provides (in anything other than dual-purpose reactors such as found in France and Great Britain) is technically inferior and more costly than would be separate dedicated facilities; (2) in the cases of all five of the acknowledged nuclear weapon states, dedicated programs rather than diversion from the civil nuclear power fuel cycle was the rule; and (3) since decisions to acquire nuclear weapons are themselves eminently political decisions, the only really effective barriers to proliferation are institutional and political, not technical.[17] Both points of view accept the theoretical linkage, but the latter, contending that the political barrier is higher than the technical barrier, rejects its probability.

Before proceeding further, it would be useful to quickly review the main features of the nuclear fuel cycle that give rise to the concerns under discussion. Because of its predominance in the commercial field and our concern with linkage between civil nuclear power and nuclear weapons, we will draw on the light water reactor fuel cycle.[18]

For convenience's sake, the nuclear fuel cycle can be divided between front-end and back-end activities with the fissioning process in a reactor serving as a link between the two. Front-end activities include mining, milling, conversion, enrichment, and fuel fabrication. Back-end activities involve storage of spent fuel, reprocessing, plutonium storage, and waste management. Broadly speaking, almost any phase of the fuel cycle is a potential trouble spot from a proliferation point of view. But more realistically, proliferation risks are greatest where directly weapons-usable materials are accessible, rather than where further processing steps are necessary to provide access to such materials. This puts emphasis on enrichment and reprocessing, and on facilities that store or use highly enriched uranium and plutonium.

Nuclear energy is the result of a chain reaction produced by the splitting, or fissioning, of uranium atoms. In a reactor, the release of nuclear energy is controlled. Without controls, the same chain reaction, admittedly with some difficulty, can be fashioned by technically skilled people with appropriate materials to result in a nuclear explosion, and hence in nuclear weapons. The relevant materials in this case are uranium 235 and plutonium.

Uranium in nature contains only 0.7 percent of the fissile isotope 235, and must be upgraded, or enriched, to between 2 and 4 percent to sustain a chain reaction and fuel a light water reactor. To be weapons-usable (that is, to be able to be exploded) however, the fissile content of uranium must be enriched significantly beyond what is required to fuel a light water reactor, typically (but not necessarily) to 90 percent, at which point the quantity required for a critical mass is sufficiently small (on the order of 10 kilograms) to be manageable for a weapons fabricator. For a country with basic enrichment capability, there is no inherent barrier (although there may be some practical difficulties) to being able to enrich to the higher levels required for nuclear explosives. Hence, both the material and the capacity to produce it are principal objects of proliferation concern.

Enrichment, however, is a complex process that, until recently, has been based almost entirely on two technologies—gaseous diffusion and ultracentrifugation—both of which have been confined to a relatively small number of countries, primarily those with nuclear weapons. That has been changing in recent years with respect to the number of states having some form of enrichment activity (research, pilot plant, or production facility) and with respect to the type of enrichment process in use. The United States, for example (because of a slowdown in the growth of nuclear power and a reduced export market), recently decided to terminate its relatively new centrifuge enrichment program that had been developed to replace the more costly and complex diffusion plants, and to pursue research and development on an atomic vapor laser isotopic separation technique as the potential next generation of enrichment technology. The newer technologies, while more cost efficient, also may have the potential for creating more serious proliferation problems because of their smaller size and the relative ease of hiding a clandestine plant from detection.[19]

The fission process that produces energy in a power reactor also produces plutonium, another fissile material that, like enriched uranium, can be used to fuel power reactors, or for explosive purposes. To be weapons usable, the plutonium must be separated from fission products and other materials by a chemical procedure known as reprocessing. Plutonium produced in a reactor typically consists of a number of different isotopes, one of which, PU 239, is particularly attractive for explosive purposes, but others of which (for example, PU 240, which accounts for about 20 percent of plutonium in a typical light water reactor) detract from the weapons-quality of the material. The longer the uranium resides in the reactor (that is, the higher the burn-up of the reactor fuel), the greater the proportion of these other isotopes, and the less attractive the produced plutonium from a weapons point of view. The reactor fuel quality of plutonium that is separated from high burn-up fuel is not adversely affected by the presence of these other isotopes. However, this is all a matter of degrees, and any separated plutonium can be used for explosive purposes (at a cost of yield and reliability, of course) and is, therefore, of proliferation concern. This is why reprocessing plants, plutonium storage facilities, and fabrication plants employing plutonium oxides in fuel assemblies (all of which involve readily accessible directly weapons-usable plutonium) are regarded as high-risk facilities from a proliferation point of view.

Conventional fuel cycle wisdom was based on the assumption that rapid growth in nuclear energy would quickly deplete economic sources of uranium, itself a finite resource, and that plutonium would become the basis for the long-term fission fuel cycle. At first, this would take the form of recycling recovered plutonium in mixed oxide fuels (uranium and plutonium), which also would provide practical experience in handling plutonium. Eventually, however, the nuclear fuel cycle would be anchored in the plutonium breeder reactor, which, because of its ability to produce more fuel than it consumed, would establish a virtually inexhaustible source of energy. Economic and political events have altered these assumptions, but they are the bases upon which the commercial nuclear age was founded.

If the decision is made not to recover the residual fuel value contained in the unburnt uranium and plutonium produced as a result of the fissioning process, one ends up with the so-called once-through uranium fuel cycle. If, on the other hand, recovery of those elements is provided for, a chemical separation or reprocessing stage is added, introducing a plutonium fuel cycle. Whether the fuel cycle is operating on a once-through or plutonium/uranium recovery basis, about one-third of the fuel in a light water reactor core is usually discharged and replaced annually. Assuming a standard size 1,000 megawatt reactor, the annually discharged fuel would contain approximately 170 kilograms of plutonium, or sufficient nuclear material for as many as 20 explosive devices the size of the bomb that destroyed Nagasaki. The actual number of devices would depend on the quality of the plutonium and the sophistication of the bomb design.

It cannot be denied that peaceful and military uses of nuclear energy share a common technological root, but possession of a civil, nuclear-powered electricity system does not necessarily lead to the acquisition of nuclear weapons. Conversely, the lack of a civil nuclear power program alone will not prevent a moderately industrialized state from making nuclear weapons. Indeed, all five of the acknowledged nuclear weapon nations acquired weapons before building civilian nuclear power.[20]

Countries with varying backgrounds and experiences, of course will see things differently. The United States, for example, is the only country to have dropped atomic bombs on cities, having developed nuclear technology as a military enterprise in the heat of war. Now its present international political and security interests could be severely compromised by the emergence of more nuclear weapon states. Therefore, the United States is more sensitive about the risks of technological abuse than are countries having fewer political or security concerns but substantial energy security problems. Such states may argue that it is countries, not technologies, that are dangerous, and that with effective controls, countries with good nonproliferation credentials have the right to make the most of nuclear energy. A state with limited security horizons, no global political commitments, and modest interests or stakes outside its region is also less likely to fear the production of nuclear weapons in some remote part of the globe. In short, the two faces of nuclear technology are likely to be seen quite differently, according to the extent of a country's global or widely cast political interests or security commitments.

What emerges from review of the linkage issue is that while the civil nuclear fuel cycle may not be the preferred route to nuclear weapons, peaceful nuclear programs can provide a means of acquiring the necessary fissile material for nuclear explosives. The Indian experience is a case in point. India's research and development program, which was allegedly related to civilian nuclear power, nevertheless was a vehicle for acquiring plutonium for a nuclear explosive device. That conclusion is not diminished either by the fact that India used a research reactor rather than a power reactor to produce the material, or that the government insists that the event involved a peaceful nuclear explosion. Nuclear materials are not exclusively either military or peaceful. As we have seen, they can serve dual ends. In the last analysis, their characterization depends upon how they are used. Meeting the challenge posed by the dual nature of nuclear technology and the interest of many nations in having access to the beneficial, peaceful uses of atomic energy without increasing the risk of weapons proliferation is the mission of the nonproliferation regime.

THE NONPROLIFERATION REGIME: ASSUMPTIONS AND EVOLUTION

As mentioned earlier, the nuclear nonproliferation regime consists of a collection of principles, understandings, agreements, procedures, and treaties featuring formal pledges not to acquire nuclear weapons or explosives that are verified by international inspection and other means—that is, by safeguards. The regime is dynamic, not static, and is the product of an evolution that began in 1945. It is fully operative today, although one key part—the NPT—effectively will end in 1995 unless extended. Eight years is not a long time in world affairs.

Underlying Assumptions of the Regime

The nonproliferation regime rests on four assumptions that matured at different points in time, but that were implicit at least to some degree from the outset of the nuclear age.

First is the belief that the spread of nuclear weapons rather than enhancing national security, is likely to lead to increased instability and reduced security. The importance of this belief cannot be overstated. It is reflected in the number of adherents to the NPT who agree that national security is better served by forswearing nuclear weapons. It provides a foundation for arguing the nonlegitimacy of nuclear weapons acquisition, and it surely has served to impede proliferation.

Second is the assumption that nuclear energy has a positive and important role to play in national energy development and security.

Third, despite the dual nature of some nuclear technologies, it is possible to establish appropriate rules and conditions for cooperation that can facilitate access to nuclear energy without unacceptably increasing the risk of proliferation. This belief is most readily identified with Atoms for Peace, but it also found expression in the initial postwar effort to establish novel international arrangements for the conduct of nuclear activity. In recent years, some uncertainties regarding the efficacy of particular rules and conditions have caused difficulties for this proposition, but it remains a central notion of the regime.

Fourth (and this is somewhat more debatable), there is a linkage between the spread of nuclear weapons (horizontal proliferation) and control of nuclear arms by the nuclear weapon states (vertical proliferation). Failure to achieve progress on the latter front can have damaging effects on efforts to contain the spread of nuclear weapons.[21] This belief took on particular prominence in negotiations on the NPT when non-nuclear weapon states sought reciprocal commitments from the weapon states in conjunction with their own agreements not to acquire nuclear weapons. It is especially strong among the neutral and nonaligned states, but in recent years has found adherents also among those living under the protection of the U.S. nuclear umbrella.[22]

These propositions are widely but not universally shared. To the extent that they are not, the international nonproliferation regime is weakened. Accordingly, any fundamental change in support for them by members of the regime or by important nonmembers can profoundly affect its very existence. The absence of universality, however, is not fatal to the existence or relevance of the regime.

Evolution of the Regime

There have been three major phases in the evolution of the nonproliferation regime:

1. An initial period of secrecy (1946–1953) that ended with President Eisenhower's "Atoms for Peace" program;
2. Atoms for Peace (1954–1974), a period of open and basically unrestricted development of peaceful nuclear energy, which ended in its original form with the Indian explosion; and
3. Today's post Atoms for Peace era, which is still changing.[23]

The transitions from one phase to another are the consequence of many factors. These include the success of various nonproliferation policies; the impact of technological developments; competitive commercial considerations; and changes in the economic, technological, and political environment in which nuclear technology is applied. Some examples include changes in energy supply and demand, the attractiveness of alternatives to nuclear energy, and the emergence from colonialism of new states that, although they lack a significant scientific-technological base, are strongly assertive about their sovereignty and convinced of their need for advanced technologies. These factors sometimes have been mutually reinforcing and sometimes contradictory. All have played important roles in shaping the characteristics of the nonproliferation regime, and they bear significantly on how effectively the IAEA is able to play its part in nuclear development, cooperation, and control. What follows is intended as an overview, not as an in-depth statement of the major developmental phases of the regime. Subsequent chapters will provide further elaboration, as required, to understand the role and activities of the IAEA.

Phase 1: 1946–1954

The approach adopted to meet the challenge of the nuclear age in the immediate aftermath of the Second World War confirms how sensitive policymakers were to the problem of dealing with a Janus-faced technology. This period was dominated by restrictive agreements and audacious initiatives, all apparent in the Trilateral Agreed Declaration of Canada, the United Kingdom, and the United States[24] and in the Baruch Plan.[25] The trilateral declaration, noting that the military use of atomic energy depended to a substantial degree on the same methods and processes that would be required for industrial uses, but recognizing the potential benefits to humankind of atomic energy, committed the three governments to withhold atomic information for industrial purposes until effective and enforceable safeguards against its use for destructive purposes had been established. At the same time, the declaration acknowledged that "no system of safeguards that can be devised will *of itself* provide an effective guarantee against the production of atomic weapons bent on aggression," and that the only complete protection against such a situation lay in the prevention of war.[26] It concluded that avoiding the destructive uses of atomic energy and promoting its peaceful and humanitarian uses was, in the final analysis, a responsibility of the entire international community acting collectively; and it called upon the United Nations to establish a commission to make specific proposals for dealing with the nuclear challenge. In these few simple observations, the declaration identified the key parameters that would dominate the nuclear debate for the next four decades.

The Baruch Plan was the United States's bold and far-reaching, but ill-fated, initiative presented to the United Nations. It was based on a review of nuclear issues prepared for the secretary of state, the

Acheson-Lilienthal Report, which had concluded that because the ability to produce special nuclear material was a critical step toward nuclear weapons, a "system of inspection superimposed on *an otherwise uncontrolled exploitation of atomic energy by national governments* will not be an adequate safeguard," and that international safeguards alone could not therefore assure effective separation of civil and military uses of nuclear energy.[27] The U.S. solution to this problem, as stated in the Baruch Plan, was to establish an international agency that would have (1) managerial control or ownership of all atomic energy activities potentially dangerous to world security (that is, all militarily relevant nuclear operations); (2) authority to control, inspect, and license all other atomic energy activities; and (3) responsibility to engage in atomic energy research and development.[28]

Opposed by the Soviet Union, which was unwilling to consider such international management and control unless and until the United States first destroyed its own nuclear weapons stockpile, the Baruch Plan failed to achieve the necessary support. Even before the plan was finally shelved, the United States, in the Atomic Energy Act of 1946, established a policy of secrecy and denial, prohibiting any peaceful nuclear cooperation until Congress was satisfied that effective international safeguards were in place.[29]

By this means and through steps taken by the Allies to buy up all existing uranium supplies, it was hoped that the uncontrolled spread of nuclear technology and materials would be prevented. Aspects of this first phase that are relevant to the establishment of the IAEA a decade later are explored in chapter 2.[30]

Phase 2: 1954–1974

The limitations of denial were soon revealed by successful Soviet fission and thermonuclear tests in 1949 and 1953 and by British testing of a nuclear explosive device in 1952. Furthermore, the United States, which needed to import uranium for its weapons production from Canada and Belgium (Belgian Congo), granted those countries some nuclear concessions.[31]

Nuclear energy represented the highest technology of the time, and national nuclear programs were moving forward in a number of countries, including France, Switzerland, and Sweden. These three nations and others had begun an inexorable march toward an independent nuclear power capability that was not subject to any international control.

The ineffectiveness of denial as a policy to prevent the spread of nuclear information and technology, let alone the development of nuclear weapons, and the acceleration of nuclear interest and activity abroad, led the United States to change its policy from denial to controlled nuclear assistance and cooperation. This important United States initiative was put forward in President Eisenhower's Atoms for Peace proposal of December 8, 1953.[32] It signalled the opening of a major new era in nuclear cooperation—an era marked by the establishment of the International Atomic Energy Agency (1957) with its soon-to-develop international safeguards system, the Treaty for the Prohibition of Nuclear Weapons in Latin America, known as the Treaty of Tlatelolco (1967), and the Treaty on the Non-Proliferation of Nuclear Weapons (1968).[33]

Atoms for Peace was more than a U.S. acquiescence to global trends of nuclear technology dispersion, although the erosion of American technological monopoly was to play an important part in defining the scope and character of arrangements for subsequent international cooperation. It also reflected President Eisenhower's desire to promote the peaceful use of atoms in order to disperse the image of the bomb that heretofore had been associated with U.S. nuclear policy. In the longer view, Atoms for Peace was also visualized as a possible first step in the direction of nuclear arms control, a start toward reducing the nuclear stockpiles of the weapon states without any of the cumbersome requirements for effective international inspection and control that have complicated future arms control talks.[34]

It would be difficult to overestimate the importance of Atoms for Peace, whether one considers it a brilliant and imaginative initiative, or premature and based on naïve assumptions about human behavior and international politics.[35] As seen by a leading foreign participant in international nuclear affairs, Atoms for Peace was a major watershed entailing "the passage from the depressing era of denials to the euphoric period of transfers, from the nuclear Middle Ages to the nuclear Renaissance."[36] Based on the initiative, the United States amended its Atomic Energy Act in 1954 to permit peaceful international nuclear cooperation. This, in turn, led to the establishment of a substantial number of bilateral agreements for cooperation (22 by the end of 1958), setting a pattern (including bilateral safeguards) soon to be followed by other nations such as Canada, the United Kingdom, and France. These agreements embraced information exchange, education and training programs, and assistance in procuring necessary equipment, including research reactors, toward which end the United States, for its agreements, provided a $350,000 subsidy. Many of these reactors had no military significance, although two, the CIRUS reactor supplied by Canada to India (along with U.S. heavy water) and the Dimona reactor, provided to Israel by France, were larger facilities—capable of producing militarily significant quantities of plutonium.[37]

During the Atoms for Peace era, two important conferences on the peaceful uses of atomic energy were held in Geneva in 1955 and 1958. These conferences opened a floodgate of technical and scientific information about virtually every aspect of the civil nuclear fuel cycle with the exception of uranium enrichment, over which the United States still held a monopoly. Another hallmark of the era, of course, was the establishment of the IAEA, which not only institutionalized global dissemination of nuclear science and technical assistance, but also evolved an internationally acceptable system of safeguards. This latter feature is central to nonproliferation, and in the view of many, could not have achieved the level of acceptance and support it enjoys in the absence of an initiative of the scope and generosity of Atoms for Peace.

Phase 3: 1974 to the Present

Nineteen seventy-four was a watershed year: several events combined to raise public concern about the adequacy of the regime to effectively contain proliferation. India's explosion of a so-called peaceful nuclear device was the most dramatic of these events. Another was the projected transfer of sensitive fuel cycle technology by France and the Federal Republic of Germany to several countries with only incipient nuclear energy programs. Three of these countries—Pakistan, South Korea, and Taiwan—were located in unstable regions, and they themselves either harbored or were targets of revanchist sentiments. With these events, the role and effectiveness of international safeguards became a central issue.

Over the course of the next several years, the United States launched a two-pronged effort to reinforce the regime and to restore confidence in controlled nuclear cooperation. First, under both the Ford and Carter administrations, an attempt was made to mobilize the principal nuclear suppliers to reach voluntary agreement on the terms and conditions under which future nuclear exports would be made. A principal objective was to ensure that safeguards would not become bargaining chips in commercial nuclear competition. Another purpose was to reduce the burden on safeguards and the regime by interposing barriers to the transfer of sensitive technologies and facilities in areas where safeguards were least likely to be effective.

The United States preferred mandatory, full-scope (comprehensive) safeguards and an embargo on sensitive nuclear transfers. Neither goal was fully attained because other suppliers balked at the idea of private agreements to impose such stringent measures on themselves and on Third World countries that had not been integrated into the decision-making process or even consulted.

However well-intentioned these initiatives in the interest of reinforcing nonproliferation and preserving the distinction between peaceful and military nuclear fuel cycle activities, they also contributed to the growing sense of uncertainty in the international nuclear arena, raising concerns about reliability and security of nuclear supply and the credibility of supplier state commitments. They also reinforced underlying concerns about discrimination, especially between developed and Third World countries.

Second, an internal U.S. decision was made by President Ford on October 28, 1976 (and reaffirmed by President Carter on April 7, 1977) to reassess conventional, nuclear fuel cycle assumptions, in particular those related to the development of spent-fuel reprocessing, plutonium recycling, and the breeder reactor. The Carter Administration strongly encouraged others to join in this reassessment, and Congress, in enacting the Nuclear Nonproliferation Act of 1978, forced the issue in several respects. Through his initiative of the International Nuclear Fuel Cycle Evaluation (INFCE), President Carter succeeded in internationalizing the reassessment approach and, at the same time, broadened it to include all aspects of the fuel cycle. INFCE did not alter basic national views about the energy value of using plutonium fuels, or cause a reversal of already adopted national policies, but it substantially raised international consciousness about the proliferation risks associated with different fuel cycle decisions, the importance of strengthening international safeguards, and the usefulness of continued exploration of supplementary institutional arrangements to reinforce the existing nonproliferation regime. It also helped to somewhat ameliorate the anxieties of a number of U.S. nuclear partners. Nevertheless, U.S. policy, on the whole, put considerable strain on international nuclear relations, and many of its legislative and executive initiatives were seen by other nations, largely its industrial nuclear partners in this instance, as threats to their own nuclear programs.

This phase is still evolving. It has brought additional nonproliferation mechanisms such as the nuclear supplier guidelines, and it has focused greater attention on the strengths and weaknesses of international safeguards and on the importance of further improving them. But, as noted above, it also has seen an erosion of confidence by some states in a regime based on pledges and international safeguards alone. Still others have indicated reduced confidence that agreements for cooperation will be fulfilled according to their original terms. Moreover, it has heightened developing countries' sensitivity to discrimination, making even more challenging the task of maintaining a broad-based nonproliferation consensus while keeping in check risky transfers of sensitive material, equipment, and technologies. We will return to a closer examination of some of these events in chapter 6.

MAJOR ELEMENTS OF THE REGIME

As noted, the elements of the nonproliferation regime include the following:

- A general world consensus and predisposition against the further spread of nuclear weapons;
- Peaceful use undertakings for nuclear cooperation and trade (normally contained in bilateral agreements for cooperation, or in multilateral treaties, or both);
- Voluntary constraints by nuclear suppliers;
- Two treaties—Tlatelolco and NPT;
- Verification of peaceful use and no-weapons pledges; and
- An international agency, the IAEA, to do the verification by safeguards and to facilitate access to peaceful uses of nuclear energy.

A Predisposition Against Nuclear Weapons

A substantial number of important countries that are in a position to do so, have elected not to acquire nuclear weapons. For one reason or another, they have concluded that their national security interests are better served by eschewing nuclear weapons. They have expressed this conviction by ratifying the NPT, thereby formally undertaking not to acquire nuclear weapons and to accept international verification of that pledge. In a few instances, ratification has come as a result of strong urging and pressure on the part of one or another of the superpowers. States in this position may feel that withdrawal from the NPT would be less damaging to their real interests than would other states.

Nevertheless, the broad consensus against the further spread of nuclear weapons reinforces nonproliferation and raises important political barriers against the legitimacy of proliferation on the part of any state. Wider support for the nonproliferation consensus increases the potential political costs for those who would attempt to cross that barrier; as a consequence, the likelihood increases that they will resist impulses to do so. A central challenge to the nonproliferation regime is to maintain this predisposition against nuclear weapons, which requires balanced and sensitive policies toward global security and stability on the part of the world's great powers, and a concomitant willingness to enforce it with sanctions.

Peaceful Use Undertakings

Given the nuclear predominance of the United States in the early years of Atoms for Peace, its arrangements for international cooperation set the general standard. Although not all suppliers followed the U.S. lead precisely (Canada, for example, provided India with a research reactor without requiring safeguards), so few others were involved that such a generalization can be made. Atoms for Peace, as mentioned, led to rewriting the restrictive U.S. Atomic Energy Act of 1946. The resulting 1954 act encouraged nuclear cooperative agreements with other nations and specified certain conditions for these agreements. Among them was an assurance by the recipient that no material or equipment transferred under the agreement and no subsequently produced material would be used for nuclear weapons or for any other military purpose; and that the agreement would be subject to mutually agreed upon safeguards. Recipients also had to agree that any retransfer by them of facilities, material, or technology would take place only according to terms specified in the original agreement; and that no transfers would be made to unauthorized parties. These undertakings involved mainly bilateral commitments. Fifteen years later the NPT would go beyond the 1954 act and provide the basis for multilateral commitments that expanded coverage for all indigenously developed nuclear materials and facilities as well.

Although they are not treaties under U.S. law, the undertakings in the agreements for cooperation constitute formal commitments among sovereign nations. Despite uncertainties and misgivings about the long-run intentions of some governments, no such undertaking has ever been shown to have been violated. The only gray area case in question is India's already mentioned 1974 test of a nuclear device that used plutonium produced in a reactor built with Canadian assistance and U.S. heavy water, subject to a peaceful use undertaking. This case did not, however, involve an explicit violation of safeguards.

India contended at the time (and continues to do so) that its actions were not precluded under its bilateral agreement, since the explosion was conducted for peaceful purposes. Canada sharply disagreed that it had entered into an agreement that permitted any kind of explosion, and challenged India's right to unilaterally interpret what may have been ambiguous language. This, the only verified incident of its kind, has not been repeated, although there is much speculation about a possible South African nuclear test based on a mysterious flash picked up by satellite in the South Atlantic in 1979. The principle of no explosions and peaceful use only remains a firmly set keystone in the arch of the nonproliferation regime.[38]

Voluntary Restraints by Nuclear Suppliers

The nonproliferation regime consists of obligatory and voluntary elements. Among the latter, one of the most significant is the body of nuclear export guidelines agreed to during a series of secret meetings in London, beginning in 1975 at the initiative of the United States. The guidelines were intended to establish uniform standards for nuclear exports to any state not having nuclear weapons, whether or not it was party to the NPT. The two most important features of the guidelines are the agreement that specified exports would require IAEA safeguards and that exporters would exercise restraint in the transfer of weapons-usable material, sensitive facilities, or technologies (that is, reprocessing, enrichment, and heavy water production); an agreed list of items whose transfer would "trigger" IAEA safeguards also was adopted. We have already noted that the United States failed to achieve its objective of full-scope (that is, comprehensive) safeguards as a mandatory condition for the export of any item on the agreed trigger list, or of the obligatory foreclosing of further exports of sensitive technologies or facilities.

These limitations notwithstanding, the guidelines are notable for several reasons. First, France—both a major supplier and a nonsignatory to the NPT—participated in the meetings and agreed to the guidelines. This marked the first instance of French cooperation with the major NPT states on terms and conditions of nuclear supply. Second, at French initiative, the guidelines extended the application of IAEA safeguards for the first time to transferred sensitive technology with a conclusive presumption that, for a period of twenty years following the initial transfer, reproduction of any facility of the type originally supplied involved transferred technology and, therefore, was itself subject to IAEA safeguards. Third, to a greater degree than any previous agreements among suppliers (such as the Zangger

Committee's safeguards trigger list that was developed to implement NPT obligations), the London guidelines contain explicit provisions requiring physical protection of transferred items and guarantees against the use of any such items for a nuclear explosive device.

Although France and the Federal Republic of Germany both resisted the idea of a multilateral mandatory restraint on further transfers of sensitive technologies and facilities, both independently announced their intention to embargo future sales of plutonium reprocessing facilities. Neither, however, would apply this new policy retroactively to agreements already made with Pakistan and Brazil. Nevertheless, the London process on the whole yielded significant positive results from the viewpoint of nonproliferation consensus among suppliers. It did so, however, at the cost of alienating developing states who were not brought into the process and who viewed the agreed rules as an attempt to impede their access to peaceful nuclear technology already in the hands of the industrial states. This sense of alienation reinforced Third World determination, first manifested in the IAEA in 1976, to take a more unified and activist role.

The Treaty for the Prohibition of Nuclear Weapons in Latin America (Treaty of Tlatelolco)

The Treaty of Tlatelolco establishes a regional nuclear-weapons-free zone. Nineteen of the twenty-three Latin American states are full parties to the treaty, which came into force in 1969. Three of the four remaining states have signed the treaty and two (Brazil and Chile) have ratified it. Their ratification, however, is conditioned upon adherence by all Latin American states. Argentina has signed but not ratified.[39] An Argentinian ratification would lift the condition being applied by Brazil and Chile and leave Cuba as the only holdout. To date, Cuba has not signed the treaty, and there is little expectation that it will do so in the near future.

Tlatelolco also contains two protocols binding consenting extra-regional states to the provisions of the treaty. Protocol I provides that all states with territorial interests in Latin America will keep their possessions free of nuclear weapons. Protocol II binds nuclear weapon states to respect the treaty, not contribute to any violation of its basic obligations, and "not to use or threaten to use" nuclear weapons against any of the contracting parties. With the exception of France, which has not ratified Protocol I, both protocols now have been ratified by all relevant states, including the United States.

Tlatelolco is an important companion to the NPT. It does not bear the burden of discrimination associated with the latter, thus avoiding outright rejection by such states as Argentina and Brazil, and it can serve as a functional equivalent to the NPT for states such as Colombia, which thus far have not joined the NPT. It also provides for the application of IAEA safeguards on all peaceful nuclear facilities, as does the NPT.

However, Tlatelolco raises some questions in one area. Although the parties pledge not to develop, test, or import nuclear weapons or to permit foreign nuclear bases to be established on their territory, Argentina contends that the treaty would permit explosions for peaceful purposes. Argentina, however, declared voluntary abstention from any such activity. Most other parties to the treaty—as well as the United States and the Soviet Union (as expressed on their ratification of Protocol II)—take the position that peaceful nuclear explosions are not permitted under the terms of the treaty.[40]

Resolution of this difference of interpretation and adherence of the few remaining holdouts would make Tlatelolco an even more significant reinforcement of nonproliferation than it already is. Even in its present status, it is an important factor in the nonproliferation regime: despite differences of interpretation about the legality of peaceful nuclear explosions, it outlaws outright acquisition of nuclear weapons, and provides an example that might be adaptable to regional agreements elsewhere.

The Treaty on the Non-Proliferation of Nuclear Weapons (NPT)

The NPT is the central juridical instrument of the nonproliferation regime. It also involves (along with Tlatelolco) a unique experiment in international safeguards to verify that nuclear materials and facilities are not used to make nuclear weapons.

Basic Provisions

The NPT states an international consensus against the further spread of nuclear weapons, and with respect to nuclear materials in peaceful use, provides for verification of the pledges of no nuclear weapons by non-nuclear weapon states. Article I states that weapons state signatories are committed not to transfer nuclear weapons or to assist non-nuclear weapon states in acquiring such weapons. Article II states that non-nuclear weapon states commit themselves not to acquire nuclear weapons or explosive devices, whereas Article III requires them to submit *all* of their peaceful nuclear activities to IAEA safeguards to facilitate verification that they are meeting their treaty obligations. All parties are committed to require IAEA safeguards on their nuclear exports, a requirement that did not exist under the IAEA statute. The NPT also contains negative and positive assurances with regard to peaceful nuclear development. The treaty stipulates that there should be no discrimination among parties with respect to such development. To this end, according to Article IV, all parties undertake to facilitate "the fullest possible exchange" of equipment, materials, and technological information."

Closing a Loophole

The NPT closed an important loophole in the earlier nonproliferation regime by obtaining nonproliferation undertakings and requiring safeguards on all nuclear activity in signatory non-nuclear weapon states. Before the NPT, U.S. and other bilateral agreements generally obliged other states only to agree not to appropriate peaceful nuclear assistance to military ends. Military efforts parallel to assisted peaceful programs, even efforts drawing on the experience or technological information provided by international cooperation and exchange, were not precluded.

Limitations

Viewed with the benefit of fifteen years hindsight, the NPT, for all it has done to stop proliferation, has some limitations and imperfections that have attracted attention since the mid-1970s. Some of the flaws are in the treaty itself, whereas others are the result of changes in the international environment in which the treaty is applied. The distinctions are not always clear.

Incomplete membership. As noted earlier, the NPT itself is not universally supported. A number of important non-weapon states are not parties to the treaty, among them India, Pakistan, Israel, South Africa, Agrentina, Spain, and Brazil. Two nuclear weapon states, France and China, also are not parties. France has said it will act as though it were a treaty party,[41] and has thus far done so, and China has made similar but less precise statements.[42] But India, Brazil, and Argentina remain strong critics and opponents. Moreover, certain states that have adhered to the treaty have become suspect regarding their intentions (for example, Libya, Iraq, and Iran).

Unfortunate development of peaceful nuclear explosive programs. From the late 1950s, the United States, joined shortly thereafter by the Soviet Union, unlatched a nuclear Pandora's box, with the development and testing of so-called peaceful applications of nuclear explosives, (PNEs). The idea of legitimate use of nuclear explosives was seized upon by India to justify its nuclear test of 1974. While the United States has since abandoned such peaceful applications and opposes their use, the Soviet Union has been continuing some experimental activity, although it allegedly is prepared to forgo them in the context of a comprehensive test ban (CTB) agreement. Be that as it may, the early U.S. and Soviet interest in peaceful nuclear explosions influenced negotiations of the NPT, and resulted in the incorporation (Article V) of provisions for ensuring that the potential benefits of peaceful nuclear explosions would be made available to non-nuclear weapon state parties by weapon state parties. No such request ever has been made.

Nonproscribed military activities. A related limitation of NPT is that although it forecloses acquisition of nuclear explosives by non-nuclear weapon states, it leaves open the possibility for them to pursue nonexplosive military applications such as submarine propulsion. In principle, the latter could take place outside safeguards if it involved indigenous material or material from a supplier that did not preclude its use for nonexplosive military purposes. This creates the possibility that a nonexplosive military program might be a cover for explosives development, or an overt complement to a covert weapons program. Additionally, there is nothing in the NPT to prevent non-weapon state parties from taking preparatory steps toward developing nuclear weapons as long as they do not actually acquire them. This, as we shall see later, has contributed to a debate over where proliferation actually begins.

Easy withdrawal from treaty obligations. Considering the critical nature of the NPT's no-weapons pledges, it comes as a surprise to some that Article X of the treaty permits a party to withdraw upon ninety days' notice if "extraordinary events related to the subject matter of the Treaty have jeopardized the supreme national interests" of the country giving notice. This provision, however, is by no means the unqualified right of withdrawal, which critics often contend the treaty allows. The requirement of justification interposes a political hurdle that a state contemplating withdrawal may find a strong deterrent to exercising its "rights."

It remains a cause for concern, however, since some countries might be tempted to withdraw because they do not perceive their security interests as being affected one way or another by such action, or out of feelings of frustration about increased restraints on the part of nuclear suppliers. Others may do so because they have what they want. Nevertheless, no state has yet invoked the withdrawal provision (thereby expressing tacit support *for* the treaty, although withdrawal is legally possible), and there is good reason to think that absent a general breakdown of confidence in the regime, or a weakening of the international consensus against proliferation, withdrawal would be politically and diplomatically difficult. The advanced nuclear states (weapon and non-weapon), which have a high stake in the continued viability of the nonproliferation regime, have in their hands the necessary political and other resources to make maintenance and reinforcement of the nonproliferation regime a matter of common concern. The central question is whether they have the political will to actively implement that concern.[43]

Binding period too short. Again, considering the importance now attached to the NPT, it seems odd in retrospect that it has a term of only twenty-five years. The U.S.-Soviet draft treaty of 1967 had provided for unlimited duration, but this was not favored by a number of advanced industrial states. In 1995, an international conference will be convened to decide "whether the Treaty shall continue in force indefinitely, or shall be extended for an additional fixed period or periods."[44] That decision will be taken by a majority vote of the parties.

The record of the IAEA and its safeguards can be expected to influence the proceedings and outcome of that conference, along with debate over nuclear arms control and questions of nuclear technology transfers and assured supply. More importantly, the future of safeguards as we know them may well depend on the fate of the NPT, because it is doubtful that safeguards would maintain adequate credibility or confidence if all facilities were not covered at least in most countries, which is the situation that exists under the NPT today.[45]

Narrow definition of proliferation. The NPT defines proliferation as the acquisition or testing of a nuclear weapon or nuclear explosive device, but is silent on acquisition of nuclear materials that could readily be diverted to make nuclear weapons—that is, highly enriched uranium and plutonium. This was no oversight, but the intended purpose of the drafters of the treaty. Testifying before the Senate Foreign Relations Committee on the NPT in 1968, William C. Foster, director of the Arms Control and Disarmament Agency noted that he had been asked by other delegations what would constitute the "manufacture" of a nuclear weapon or other nuclear explosive device under Article II of the treaty. He reported that the United States had stated that while it would be difficult to formulate a comprehensive definition or interpretation, "Neither uranium enrichment nor the stockpiling of fissionable material in connection with a peaceful program would violate Article II so long as these activities were safeguarded under Article III. Also clearly permitted would be the development, under safeguards, of plutonium-fueled power reactors, including research on the properties of metallic plutonium. . . ."[46]

For its part, the Federal Republic of Germany, in depositing its signature of the NPT with the United States in 1969, went to great lengths to emphasize its understanding that no nuclear activities for peaceful purposes were to be prohibited, and that cooperation even in sensitive activities could not be denied merely because they theoretically could relate to the manufacture of nuclear weapons or explosive devices.[47] Thus, it seems highly improbable that the treaty could have been concluded if it had limited commercial production and use for civil purposes of weapons usable material.

As we have seen, however, by the mid-1970s, altered international circumstances led many analysts and policymakers, especially in the United States, to question the adequacy of the nonproliferation regime, and whether the definition of proliferation should not take into account capabilities as well as behavior. It was felt that some states may have joined the treaty to gain access to nuclear technology and to get close to acquiring weapons-usable material, without necessarily appearing to be doing so, for a military purpose. Faced with the prospect that some countries might get too close to nuclear weapons, the United States took the lead in trying to limit, if not foreclose, transfers of sensitive nuclear technology and facilities. This, more than any of the other problems discussed, reflects a case of external changes rather than the discovery of hidden faults resulting in perceived limitations in the NPT. IAEA safeguards also felt the impact of these external changes.

Discrimination. A common complaint voiced about the NPT is that it is discriminatory. To the extent that the NPT distinguishes between nuclear weapon states (defined as those that exploded a nuclear device before January 1, 1967) and non-nuclear weapon states, and imposes different obligations on the two categories, it is discriminatory. There are, however, obligations on the weapon states as well, including Article IV mentioned earlier, and Article VI, which imposes on member states the obligation "to pursue negotiations in good faith" on nuclear arms control and disarmament measures. Two successive international review conferences on the NPT, held in 1975 and 1980, have witnessed general dissatisfaction with the lack of progress made thus far by the superpowers. Indeed, there have been rumors of possible defections resulting from frustration or dissatisfaction, but fortunately, none have occurred, and although the issue was prominent at the 1985 NPT Review Conference, it was less threatening and disruptive than anticipated. However, in light of the fact that in 1995 the parties will have to decide on the fate of the treaty, it is necessary to head off any potential defections. For not only could the NPT be endangered, it could have no future.

Safeguards. The concept of safeguards is fundamental to nonproliferation, peaceful nuclear cooperation, and trade. Consequently, verification of pledges by non-nuclear weapons states not to use their nuclear activities to make nuclear weapons or explosives, and to ensure, by inspection, that nuclear materials in peaceful use are accounted for, is central to the basic proposition of Atoms for Peace that the civil and military uses of nuclear energy can be separated. The willingness to accept and assist national development of nuclear energy, so long as it was done openly and subject to nonproliferation commitments, marked a radical departure from the official American thinking of 1946, when the Acheson-Lilienthal Report advised that safeguards alone would not be adequate to achieve the objective of preventing the spread of nuclear weapons. At that time, the report and the subsequent Baruch Plan favored the more far-reaching solution of international ownership and control in lieu of safeguarded national development of nuclear energy. Safeguards at that time were conceived *not* as verification mechanisms of the type applied by the IAEA today, but as more extensive measures closer in tone to domestic safeguards that typically play a preventive role and include physical security measures. In other words, the safeguards that were earlier considered to be inadequate were conceived as serving a broader set of responsibilities than current international safeguards. Confusion over this issue is one of the sources of the safeguards problem today.

Safeguards resurfaced with Atoms for Peace, where they took on a somewhat different meaning. In presenting his proposal to the United Nations, President Eisenhower said the cooperation he proposed could "be undertaken without the irritation and material suspicions incidental to any attempt to set up a completely acceptable system of worldwide inspection and control."[48] That remark, of course, referred to his proposal that the United States and Soviet Union contribute fissile material from their weapons stockpile to an international agency for developing peaceful uses of atomic energy. As things turned out, the Atoms for Peace idea rapidly passed through several developmental stages, the concept of verification by safeguards took form, and safeguards became a prerequisite for international nuclear cooperation.

As applied by the IAEA, safeguards are a major and unique innovation. Participating sovereign states have agreed to permit inspection by an international authority to verify that nuclear materials in peaceful use are accounted for and nonproliferation commitments are being met. Even when applied bilaterally by a supplier to a recipient, as safeguards were for nearly the first decade of the Atoms for Peace era, they represented a novel and important deviation from conventional interstate relations and a significant inroad on national sovereignty.

More recently, IAEA safeguards have been challenged as ineffective. In part, this stems from misunderstandings about the nature and purpose of IAEA safeguards and its mandate. But it also derives from concerns about effectiveness of international safeguards for large bulk-handling facilities—such as reprocessing or enrichment plants—which provide direct access to weapons-usable material in relatively easily divertable forms. The challenge also reflects doubts about how the IAEA has organized to meet its safeguards responsibilities. There are questions about the ability of its management structure to ensure that an alarm will be quickly sounded if and when needed. Although these questions will be dealt with later, they have been mentioned here to introduce the key elements of the international nuclear nonproliferation regime.

An International Atomic Energy Agency

The final component of the nuclear nonproliferation regime is the IAEA. The IAEA is the organizational core of the nonproliferation regime. Its origins are to be found in President Eisenhower's Atoms for Peace initiative, and the agency appropriately can be viewed as the first major institutionalized expression of a U.S. policy to internationalize the peaceful benefits of atomic energy and to channel nuclear technology development toward constructive and nonmilitary ends.[49]

Both its supporters and critics acknowledge that it has been one of the most professional and, until recently, one of the least politically troubled international organizations in the United Nations family. The IAEA is widely credited with some exceptional contributions to the cause of peace, and with facilitating broad access to a technology that, for all of its problems, continues to be regarded by many as a potentially important resource for global socioeconomic development and energy security. Nonetheless, the agency has not lacked criticism. Some claim the IAEA has gone too far in promoting a technology that in one view is fraught with problems of safety (witness the Three Mile Island accident and the more recent Chernobyl tragedy in the Soviet Union), waste management, and proliferation. Others claim it has not gone far enough in promoting nuclear power development.[50]

A Multiplicity of Roles

The IAEA plays a number of important roles. It provides a framework for interaction on nuclear issues and for elaboration and refinement of the international nuclear regime. It is also a forum for discussion, exchange of views, and the formulation of initiatives relevant to peaceful nuclear uses. Furthermore, it is a vehicle for the provision of services relative to the utilization and control of peaceful nuclear development, including implementation of decisions taken outside the agency itself. And it is a symbol of the commitment to share broadly and responsibly the peaceful benefits of nuclear energy.

A Framework for Nuclear Cooperation

The IAEA provides a framework for nuclear cooperation and, as such, its existence and activities can support nuclear decisions taken elsewhere. Bilateral agreements and national policies for nuclear cooperation and multilateral arrangements on rules of the game are formed with the agency in mind. International nuclear diplomacy presumes the existence of the IAEA, and its presence directly and indirectly affects the outcome of that diplomacy. While this cannot always be documented, it is an inescapable part of the international nuclear environment and the regime for cooperation and control.

One important aspect of this role for the IAEA is that it publicizes such policy decisions. For example, the London suppliers group secretly negotiated guidelines on nuclear export policy between 1975 and 1977. After much grumbling by states that were non-nuclear weapons suppliers, the guidelines were formalized and publicized in a series of parallel statements sent to the agency by the fifteen participating states. The agency subsequently issued an Information Circular[51] setting forth the texts transmitted by the suppliers of their intended practices. Another, less publicized example is the informal

Zangger Committee of NPT nuclear exporters. This group began meeting shortly after the NPT went into effect, for the purpose of working out implementation of NPT safeguards responsibilities,[52] and to establish among NPT suppliers an agreed list of items whose supply should trigger the IAEA safeguards.[53] Among other things, procedures such as these can be seen as reinforcing the notion of the agency as a validator of state actions, thereby reinforcing the legitimacy of both state actions and the agency itself.

A Forum for Nuclear Matters

The agency's value as a forum is perhaps best illustrated by the fact that the Soviet Union and the United States have substantially insulated the agency from the tensions between them. For many observers, the positive Soviet-American cooperation on IAEA matters is all the more remarkable given the uneven history of their general relations. Some states quietly regret this symbiosis because it diminishes opportunities for them to exploit superpower differences on behalf of their own interests—an opportunity that occurred more frequently during the first five years of the agency's existence.

Although the IAEA is not immune to politics, as we shall discuss in detail later, it—more than many other international institutions—has been better able to provide a point of contact and a place to discuss, if not to solve, common problems. As a result, it has been able to focus its attention on its mandatory responsibilities. Even with the recent intrusion of extraneous political issues into the IAEA and their inevitable debilitating effects, the agency's function as a forum remains largely intact, especially, but by no means exclusively, for East-West relations. The fact that much of this dialogue occurs in the wings of the IAEA rather than on the stage of its formal institutions does not detract from its importance, and is reflected in East-West behavior in those institutions.

As a forum, the IAEA can be the locus of negotiations and decisions related to nuclear activities, to the terms and conditions for nuclear cooperation, and to the elaboration and implementation of general principles. The agency's development of regulations for the transportation of radioactive materials and its sponsorship of a convention on physical security, as well as its increasing emphasis on nuclear safety, are examples of this role.[54] Its establishment of a committee to work out the implementation of safeguards responsibilities assigned by the NPT, which led to the adoption of the NPT safeguards document, is another.[55]

A Supplier of Technical Assistance and Services

At the broadest level, the service function of the IAEA is most succinctly defined in its statute that directs it to "accelerate and enlarge the contribution of atomic energy to peace, health, and prosperity throughout the world"; and to "ensure, as far as it is able, that assistance provided by it or at its request or under its supervision or control is not used in such a way as to further any military purpose."[56] This has given rise to an extensive network of technical assistance in uses of radioisotopes and radiation in agriculture, life and physical sciences, nuclear power, fuel cycle activity, and nuclear safety and waste management. These are further discussed in chapter 3. The agency's other major function is to administer international nuclear safeguards that implement both its statutory responsibilities and those assigned to it by the NPT in 1970. The IAEA director general, Hans Blix, has emphasized that safeguards also have a service dimension, noting that where they are applied effectively to all of a country's nuclear facilities, they serve to inspire confidence and to reduce regional tension, thereby facilitating the development of nuclear energy.[57]

A Repository and Supplier of Nuclear Materials

In his original proposal, President Eisenhower visualized an international agency as a repository to which governments—principally the United States and the Soviet Union—could "make joint contributions from their stockpiles of normal uranium and fissionable materials." In addition, he envisioned that such an agency could "devise methods whereby this fissionable material could be allocated to serve the peaceful pursuits of mankind."[58] This radical part of Atoms for Peace faded into the background as negotiations over the new agency progressed and developed in other directions.

Its statute did not, however, foreclose the agency from serving as a repository. Article IX, for example, permits members to make source and special fissionable materials available to the agency for storage in agency depots; and under Article XII, the agency has the right to "require deposit of any excess" special fissionable materials that are produced from materials that *it supplies* beyond what is required for peaceful research or reactor use. These authorities thus far remain unused.[59]

The idea of the agency as a custodian of deposits of fissile materials early gave way to the lesser concept of the agency as a broker or clearinghouse for bilateral cooperation.[60] It was this latter concept that ultimately was codified in the agency's statute although, as noted, provisions for a custodial role were retained. This change in emphasis limited the IAEA's direct responsibility for control over the distribution of nuclear material as well as its opportunity to have a substantial impact on shaping the emerging nuclear marketplace.

Safeguards and Verification

The IAEA survived an inauspicious beginning (resulting from the predominance of bilateralism in the early stages of international nuclear cooperation) to become one of the more significant international institutions of the second half of the twentieth century. Its principal, but not exclusive, importance today derives from its system of international safeguards. The IAEA developed and implemented the international safeguards system that transformed it into a central player in the international nuclear game. The safeguards system had to be created from scratch beginning in 1961. By the end of 1985, the IAEA had 163 safeguards agreements with 96 countries. Although inspections are only a part of safeguards, the acceptance of on-site inspections by international officials was unprecedented and remains singularly important today.

Putting international safeguards into operation was not easy to achieve. During negotiation of the IAEA statute, India and the Soviet Union tried to restrict incorporation of the safeguards principle. Joined by others, they were able to derail a U.S. proposal that members be obligated to accept safeguards. The Indians and Soviets preferred voluntary, not mandatory, acceptance. Hence, membership and safeguards were decoupled in the statute. Also, the statute did not oblige member states to require safeguards for their bilaterally supplied assistance. Individual suppliers could require such safeguards as a condition of supply, but it was not until the NPT that those who ratified the treaty became obligated to require IAEA safeguards on exports.[61] Early start-up of the initial circumscribed safeguards went slowly, as a result of dilatory tactics on the part of those members who remained antagonistic to the idea of international safeguards. The bilateral safeguards practices of the United States—largely developed and instituted for commercial reasons—did not help, because they removed pressure for early implementation of agency safeguards.

Several landmark events crucial to the development of IAEA safeguards deserve mention here. Three events were prima facie positive events reinforcing the concept and international acceptance of IAEA safeguards; a fourth, though initially perceived as destructive, may ultimately prove to have further reinforced international safeguards. These events are described below.

The U.S. shift to IAEA safeguards. The United States altered its domestic legislation to permit international cooperation in 1954 on the heels of the Atoms for Peace initiative. Safeguards were one of the key conditions to international cooperation and, as no international safeguards were in place, the United States deployed a bilateral safeguards system. By the time the IAEA came into being in 1957, there were agreements with more than two dozen states. Many of those agreements included a proviso for eventual transfer of safeguards to an international agency, assuming one was established. And even after the IAEA came into being, the United States concluded an agreement with the newly formed European Atomic Energy Community (EURATOM), entrusting it with safeguards on U.S. nuclear material supplied to community members. All of this helped to minimize any pressure for early deployment of IAEA safeguards once the agency was created.

In 1962, however, the United States transferred safeguards of its nuclear exports (except for EURATOM) to the IAEA and required, as a condition for new or amended cooperation agreements, the acceptance of agency safeguards. This decision galvanized the agency. Under its statute, the IAEA can apply safeguards in three circumstances: (1) when the agency is a supplier under a project agreement; (2) when voluntarily requested by a state to safeguard any of its nuclear activities; and (3) when so requested by the parties to a bilateral agreement.[62]

Agency projects have been limited in number, and (outside of obligations incurred as a result of treaty obligations) few states have made unilateral requests for safeguards.[63] As for purely unilateral submissions (that is, those not the result of a requirement imposed by a supplier as a condition of supply), only Mexico and Yugoslavia have opted for that provision. In fact, when the United States adopted its policy of placing safeguards responsibilities in the hands of the IAEA, it encountered resistance from many of its bilateral partners who preferred direct inspection by the United States to that of the agency. Japan was a notable exception. One reason for this general reluctance may have been uncertainty about how a new international bureaucracy would manage such a responsibility. Another may have been the sense that bilateral safeguards could be rationalized as just another condition for a cooperative or commercial transaction, whereas agreement to international safeguards, including inspection, was a genuine derogation of national sovereignty, an act viewed as quite serious. A third was the possible commercial and technological advantage that might accrue to some nations through espionage by their nationals on the IAEA safeguards staff.

In any event, the transfer was completed by the mid-1960s. Today, international safeguards are so firmly entrenched and accepted that it would be nearly impossible to envisage a return to bilateral safeguards by either suppliers or recipients.

A change in Soviet support. Second, and even more significant, was the termination of systematic Soviet opposition to IAEA safeguards. Signs of this shift appeared early in 1962, when the tone of Soviet participation in the IAEA's board meetings and other forums began to moderate. Soviet support became evident in 1963, at the time of the signing in Moscow of the Partial Test Ban Treaty. This shift of Soviet policy not only facilitated the agency's development of safeguards policy, it also added legitimacy to the agency's safeguards system. Moreover, it deprived other pockets of opposition of a politically important ally and effectively ended any significant opposition to the consolidation and extension of international safeguards. By 1965, the safeguards system had been reviewed, revised, and expanded, and the new document had been accepted.[64] Extension of the revised system to include facilities for spent-fuel reprocessing and fuel fabrication in 1966 and 1968, respectively, occurred after vigorous but constructive negotiations involving some important concessions by several participants. Once the hurdle of political opposition had been overcome, discussions on safeguards became more technical. The bargaining and negotiation were not over the legitimacy of the safeguards principle per se, but over how to preserve the agency's integrity while protecting legitimate national concerns. The revised system constituted the first program of international inspection mutually supported by East and West. That solildarity remains firmly in place today.

The safeguards assignment of the NPT. The third event contributing to the transformation of the IAEA into a centrally important nuclear institution was the decision to have the agency provide safeguards under the NPT. This is significant in two respects. First, it acknowledged the agency's acceptance in implementing safeguards and was a vote of confidence in its ability to undertake even weightier responsibilities with important arms control overtones. As we shall see, that judgment was confirmed at the 1985 NPT Review Conference. Second, it gave the IAEA a tremendous boost, making it the keystone of the nonproliferation regime, and catapulting it from the periphery to the center of the international political system, where its newly elevated importance would make it a more inviting political target. It also contributed to another transformation (stimulated largely by changes in the world political climate) from a primarily technical into a frankly political entity.

The Israeli attack on the Iraqi research reactor. The three events just discussed all served to move the cause of IAEA safeguards forward, but the fourth event did not. In June 1981, Israel carried out an air attack against Iraq's nearly completed, large research reactor. This action impugned the technical effectiveness of IAEA safeguards activities and the value of their findings. It also implicitly questioned the likelihood and effectiveness of any international response, should the IAEA, in the course of carrying out safeguards, detect a diversion of nuclear material. The doubts of many members of Congress, who took part in hearings on the attack, gave so much support to Israel's argument that an independent observer would have thought the agency was at fault for the attack.[65] The aftermath of this event will be discussed in detail later, but it may be noted here that in the long run the importance to the international community of effective safeguards has come to be even more substantially appreciated.

The Symbolic Role of the IAEA

The symbolic role of the agency needs little elaboration. It is the organizational expression of the commitment to extend the peaceful benefits of the atom. This cuts two ways. Those who favor the more restrictive policies for nuclear cooperation that have been practiced by major nuclear suppliers in recent years seek to legitimize those policies by linking them with the agency. An example of this effort is the IAEA's publication of the London supplier guidelines. Supporters of such policies see agency identification with them as consistent with the spirit of Atoms for Peace, whose purpose, they contend, is "responsible development," meaning, in their opinion, the development and cooperation consistent with the avoidance of problems posed by the spread of nuclear arms. However, for others, the IAEA's acquiescence in such activities diminishes rather than promotes peaceful nuclear cooperation and is tantamount to repudiation of the agency's *raison d'être*. Either way, the IAEA symbolizes an international commitment to nuclear cooperation that is consistent with the objective of nonproliferation.

The preceding discussion can leave no doubt about the extraordinary complexity of the nuclear problem and of avoiding nuclear proliferation. At the heart is a dual-use technology that can be appropriated to peaceful or military purposes according to the dictates of national political authority. That decision, in turn, will be driven by the interaction of a large number of considerations, including perceptions of security risks, assessments of national political interests, dynamics of the internal political situation, and the role that nuclear weapons or the option to acquire them can play in meeting those risks, fulfilling those interests, or satisfying those internal political dynamics. To succeed, nonproliferation must effectively address all of these concerns. Clearly, a number of them lie beyond the capacity of the IAEA and its international safeguards system to deal with alone, and in some cases even the comprehensive nonproliferation regime outlined above may not suffice to head off a determined proliferation.

Nevertheless, the IAEA and safeguards have a vital role to play in maximizing the possibility for nonproliferation to succeed. By establishing high credibility for international safeguards and the integrity of their implementation, and by earning the confidence of the international community at large and its individual members, the IAEA can help ameliorate security concerns and diminish the political attraction of acquiring nuclear weapons or maintaining the nuclear weapons option. By facilitating and ensuring access to nuclear materials and technology for legitimate peaceful nuclear purposes, the IAEA can diminish incentives for states to seek nuclear autarchy outside the framework of the international safeguards system. To achieve either goal requires the active support and cooperation of the agency's membership. That is the challenge, and that is the subject of this book.

NOTES

1. The problem of loss of confidence is largely, although not entirely, an American phenomenon. For many states, particularly those in Western Europe, some of the difficulties that have arisen in the IAEA were to be expected, given the structure of the international system and the nature of international relations.

2. Of course, safeguards are not the only, or even the principal, barrier to proliferation. But effective safeguards can contribute significantly to alleviating concerns that states may have about the nature of nuclear activities in neighboring countries, and thereby remove or reduce national incentives to acquire nuclear weapons. By the same token, safeguards are an important

mechanism by which states with nuclear programs can provide assurances to others as to the peaceful nature of their nuclear activities. In the absence of international verification safeguards, states might be less prepared to eschew nuclear weapons or the creation of a nuclear weapons option.

3. A number of the themes that will be developed in this book appeared in abbreviated form in Lawrence Scheinman, *The Nonproliferation Role of the International Atomic Energy Agency: A Critical Assessment*, (Washington, D.C., Resources for the Future, 1985).

4. There exists an extensive literature on nonproliferation that could not be adequately represented even in an extended note. What follows is a representative list of books or articles that cover most of the main developments.

An early and cogent statement of the proliferation problem is in Leonard Beaton and John Maddox, *The Spread of Nuclear Weapons* (London, Chatto and Windus, 1962).

The U.S. policy toward nuclear proliferation before the nonproliferation treaty can be found in William B. Bader, *The United States and the Spread of Nuclear Weapons* (New York, Pegasus, 1968).

How proliferation has been viewed from different national perspectives is very effectively provided in George Quester, *The Politics of Nuclear Proliferation* (Baltimore, Md., The Johns Hopkins University Press, 1973).

The breakdown of the stable relations that prevailed in the civil nuclear arena before 1974 are well analyzed in Bertrand Goldschmidt and Myron B. Kratzer, *Peaceful Nuclear Relations: A Study of the Creation and Erosion of Confidence* (New York, The Rockefeller Foundation and the Royal Institute of International Affairs, 1978).

A closer look at the dynamics of change from the U.S. perspective is taken by Michael Brenner, *Nuclear Power and Non-Proliferation: The Remaking of U.S. Policy* (Cambridge, England, & New York, Cambridge University Press, 1981).

A good overview of the interaction of the political, economic, and technological dynamics of nonproliferation is in William C. Potter, *Nuclear Power and Nonproliferation: an Interdisciplinary Perspective* (Cambridge, Mass., Oelgeschlager, Gunn & Hain, 1982).

For a statement of the Carter Administration's philosophy and strategy, see Joseph S. Nye, "Nonproliferation: a long-term strategy," *Foreign Affairs*, vol. 56 (April 1978) pp. 601–623. An inside perspective on the effectiveness of that policy is to be found in Gerard Smith and George Rathjens, "Reassessing Nuclear Nonproliferation Policy," *Foreign Affairs*, vol. 59 (Spring 1981) pp. 875–894.

An insight into some aspects of the approach adopted by the Reagan Administration is found in Lewis A. Dunn, *Controlling the Bomb: Nuclear Proliferation in the 1980s* (New Haven, Conn., Yale University Press, 1982), while an attempt to place all of the U.S. cooperation and nonproliferation policies in an evolutionary context appears in Lawrence Scheinman and Joseph Pilat, "Toward a More Reliable Supply: U.S. Nuclear Exports and Non-Proliferation Policy," Report LA-UR-86-311 (Los Alamos, Los Alamos National Laboratory, 1986).

Foreign perspectives on U.S. nonproliferation policy during the past decade are to be found in Pierre Lellouche, "International Nuclear Politics," *Foreign Affairs* vol. 58 (Winter 1979–1980) pp. 336–350; T.T. Polouse, *Nuclear Proliferation and the Third World* (New Delhi, ABC Press, 1982); John Simpson and Anthony G. McGrew, eds., *The International Nuclear Non-Proliferation System: Challenges and Choices* (New York, St. Martin's Press, 1984); Erwin Hackel, Karl Kaiser, and Pierre Lellouche, *Nuclear Policy in Europe: France, Germany and the International Debate* (Bonn, Forschungsinstitut der Deutschen Gesellschaft fur Auswartigesamt, 1981); and K. Subrahmanyam, ed., *Nuclear Proliferation and International Security* (New Delhi, Lancer International and Institute for Defense Studies and Analysis, 1985).

The most useful compendium of information on recent developments in the nuclear nonproliferation arena is Leonard S. Spector, *Nuclear Proliferation Today* (New York, Vintage Books, 1984) and Leonard S. Spector, *The New Nuclear Nations* (New York, Random House, 1985).

5. This argument has been most forcefully made by Kenneth Waltz, "The Spread of Nuclear Weapons: More May Be Better," *Adelphi Paper*, #171 (London, International Institute for Strategic Studies, 1981). See also John J. Weltman, "Nuclear Revolution and World Order," *World Politics*, vol. 32 no. 2 (January 1980) pp. 169–193.

6. Pre-proliferation activities that suggest that a country is laying the groundwork to exercise the nuclear option without actually crossing the threshold also can be destabilizing.

7. Appendix I contains a list of NPT parties as of the date of publication.

8. On Israel, see Shai Feldman, *Israeli Nuclear Deterrence* (New York, Columbia University Press, 1982). On South Africa, consult Robert Jaster, "Politics and the Afrikaner Bomb," *Orbis* (Winter 1984), and J.E. Spence, "South Africa: The Nuclear Option," *African Affairs* (October 1981). A more recent analysis is George Barrie, "South Africa" in Jozef Goldblat, ed., *Non-Proliferation: The Why and the Wherefore*, A SIPRI Study (London & Philadelphia, Taylor and Francis, 1985) pp. 151–159.

9. See Stephen Cohen, "Pakistan" in Edward A. Kolodziej and Robert Harkavy, eds., *Security Policies of Developing Countries* (Lexington, Mass., Lexington Books, 1982); Zalmay M. Khalilzad, "Pakistan," and Ashok Kapur, "Pakistan" in Jozef Goldblat, ed., *Non-Proliferation: The Why and the Wherefore* (see note 8 above) pp. 131–140 and 140–149.

10. For a discussion of India, see K. Subrahmanyam, "Regional Conflicts and Nuclear Forces," *Bulletin of the Atomic Scientists* (May 1984) pp. 16–19;

Girilal Jain, "India," and Rodney Jones "India" in Goldblat, ed., *Non-Proliferation: The Why and the Wherefore* (see note 8 above), pp. 89–100 and 101–123; Raju G.C. Thomas, "India's Nuclear and Space Programs: Defense or Development?" *World Politics* vol. 38 no. 2 (January 1986) pp. 315–342.

11. Argentina's nuclear posture is cogently discussed in Daniel Poneman, *Nuclear Power in the Developing World* (London, George Allen and Unwin, 1982) and in Douglas L. Tweedale, "Argentina" in James E. Katz and Onkar S. Marwah, eds., *Nuclear Power in Developing Countries* (Lexington, Mass., Lexington Books, 1982).

12. The view is bluntly expressed by K. Subrahmanyam, director of the Institute for Defense Studies and Analysis, New Delhi, in saying, "The Non-Proliferation Treaty is pernicious because it legitimizes the use of nuclear weapons by a few weapons powers." K. Subrahmanyam, "Regional Conflicts and Nuclear Fears," *Bulletin of the Atomic Scientists* (May 1984) p. 18.

13. As these nations enter the nuclear export market, their views will become increasingly important.

14. For a pointed discussion, see Harald Müller, "Superpowers' Unfulfilled NPT Promise," *Bulletin of the Atomic Scientists* (September 1985) pp. 18–20.

15. If the Third Non-Proliferation Treaty Review Conference of 1985 is any indication, some of these demands may be on the wane. Many Third World countries have become disillusioned about nuclear power, as distinguished from the nonpower uses of nuclear technology, because of the infrastructure and financial requirements necessary to make effective use of nuclear power technology.

16. See Albert Wohlstetter, *Swords from Ploughshares* (Chicago, University of Chicago Press, 1979); John Holdren, "Nuclear Power and Nuclear Weapons: The Connection Is Dangerous," *Bulletin of the Atomic Scientists*, vol. 31 no. 1 (Jan. 1983) pp. 40–45.

17. Bernard Spinrad, "Nuclear Power and Nuclear Weapons: The Connection Is Tenuous," *Bulletin of the Atomic Scientists*, vol. 31 no. 2 (Feb. 1983) pp. 42–47; Chauncey Starr, "Uranium Power and the Horizontal Proliferation of Nuclear Weapons," *Science*, (June 1, 1981) pp. 952–957. For a political and strategic perspective on these issues, consult Ted Greenwood, Harold A. Feiveson, and Theodore B. Taylor, *Nuclear Proliferation: Motivations, Capabilities and Strategies for Control* (New York, McGraw Hill, 1977) and Stephen M. Meyer, *The Dynamics of Nuclear Proliferation* (Chicago, University of Chicago Press, 1984).

18. Useful and readable descriptions of the technical aspects of the nuclear fuel cycle can be found in Wohlstetter, *Swords from Ploughshares* (see note 16 above), and Ford-Mitre Nuclear Policy Study Group, *Nuclear Power: Issues and Choices* (Cambridge, Mass., Ballinger Press, 1977).

19. Enrichment processes pose an additional problem for nonproliferation in that they provide a path to the bomb without a nuclear power program and thus possibly without any IAEA involvement.

20. In fact, the five nuclear weapon states all acquired bombs before building a civil nuclear power industry. France was a partial exception to this rule. See Lawrence Scheinman, *Atomic Energy Policy in France Under the Fourth Republic* (Princeton, Princeton University Press, 1965).

21. It is, however, curious that 1968 was not only the year of the NPT, but also the year of Glassboro, a step on the road to SALT I, yet they were quite separate events in the Department of State and Arms Control and Disarmament Agency bureaucracies. No linkage was made between the NPT and strategic arms control.

22. This is perhaps a more debatable fundamental proposition (as implied in note 21), especially in view of the fact that proliferation has not increased despite very limited progress in arms control. But this is nevertheless a linkage that is routinely asserted and that holds potential for triggering an unravelling of the regime even if only by way of becoming a convenient excuse for noncompliance with nonproliferation undertakings by states driven by other motives or considerations.

23. One could presumably select other breakpoints, such as 1970, the year in which the NPT came into force, and 1978, the year in which the United States Nuclear Non-Proliferation Act (NNPA) became law, but our purpose is to identify not merely specific events, but changes in orientation and behavior resulting in changes in the regime itself. Specific events, such as passage of the NNPA, often mark the culmination rather than the point of origination of change.

24. "Joint Declaration by the Heads of Government of the United States, the United Kingdom and Canada, November 15, 1945," in U.S. Department of State, Historical Office, Bureau of Public Affairs, *Documents on Disarmament, 1945–1959*, Pub. No. 7008, 2 vols. (Washington, D.C., U.S. Government Printing Office, 1960) vol. I, p. 1.

25. "The Baruch Plan: Statement by the United States Representative (Baruch) to the United Nations Atomic Energy Commission, June 14, 1946," in *Documents on Disarmament* (see note 24 above) vol. I, p. 10.

26. See note 24 above.

27. "A Report on the International Control of Atomic Energy," prepared for the Secretary of State's Committee on Atomic Energy (Acheson-Lilienthal Report); reproduced in part in *Nuclear Safeguards: A Reader*, a report prepared by the Congressional Research Service, Library of Congress, for the Subcommittee on Energy Research and Production, transmitted to the Committee on Science and Technology, U.S. House of Representatives, 98th Congress 1st Session, December 1983 (U.S. Government Printing Office) pp. 46–51.

28. See Leneice N. Wu, *The Baruch Plan: U.S. Diplomacy Enters the Nuclear Age*, a report prepared for the Subcommittee on National Security Policy and Scientific Developments of the Committee on Foreign Affairs, U.S. House of Representatives, August 1972 (Washington, D.C., U.S. Government Printing Office, 1972).

29. P.L. 79-585, 60 Stat. 755.

30. Many of the salient issues are interestingly explored in Bertrand Goldschmidt, *L'Aventure Atomique* (The Atomic Adventure) (Paris, Librarie Athenée Fayard, 1981).

31. For a discussion, see Richard G. Hewlett and Francis Duncan, *A History of the United States Atomic Energy Commission*, vol. II, *Atomic Shield, 1947–1952* (Washington, D.C., U.S. Atomic Energy Commission, 1972) pp. 479–484.

32. "United States 'Atoms for Peace' Proposal: Address by President Eisenhower to the General Assembly, December 8, 1953," in *Documents on Disarmament* (see note 24 above) vol. I, pp. 399–400.

33. On negotiating the IAEA statute, see Bernard G. Bechoefer, "Negotiating the Statute of the International Atomic Energy Agency," *International Organization* (Winter 1959). On the Treaty of Tlatelolco, see John Redick, "The Tlatelolco Regime and Nonproliferation in Latin America," in George Quester, ed., *Nuclear Proliferation: Breaking the Chain* (Madison, University of Wisconsin Press, 1981) pp. 103–134. On the Non-Proliferation Treaty, see Mason Willrich, *The Treaty on the Non-Proliferation of Nuclear Weapons* (Charlottesville, Va., Michie Press, 1979); Mohamed Ibrahim Shaker, *The Nuclear Non-Proliferation Treaty: Origin and Implementation, 1959–1979*, 3 vols. (London and New York, Oceana Publications, 1980).

34. These themes are explored in Henry Sokolski, "Eisenhower's Original Atoms for Peace Plan: The Arms Control Connection," a paper delivered at a seminar of the International Security Studies Program of the Woodrow Wilson International Center for Scholars, July 1983. *Occasional paper #52.*

35. This author believes that on balance it was more helpful than harmful because it set boundaries and created expectations regarding a technology whose spread could not be forever contained. Controlled cooperation, even with limits, is to be preferred to no control over a high-risk technology whose dissemination cannot be foreclosed.

36. Bertrand Goldschmidt, "From Nuclear Middle Ages to Nuclear Renaissance," in Joseph F. Pilat, Robert F. Pendley, and Charles K. Ebinger, eds., *Atoms for Peace: An Analysis After Thirty Years* (Boulder, Colo. and London, Westview Press, 1985) p. 112.

37. Reactors employing quantities of highly enriched uranium, although producing minimal quantities of plutonium, of course represented a proliferation risk unless the highly enriched uranium was pre-irradiated by the supplier as sometimes is the case. In any event, starting in the late 1970s the United States began a program of substitution of medium- and low-enriched uranium fuels for higher enriched forms.

38. There is, of course, a rich body of speculation about uranium and highly enriched uranium losses that might have ended up in Israel.

39. See John Redick, "The Tlatelolco Regime and Nonproliferation in Latin America" (see note 33 above). See also, Jorge A. Aja Espil, "Argentina," and Jose Goldemberg, "Brazil" in Jozef Goldblat, ed., *Non-proliferation: The Why and the Wherefore* (see note 8 above) pp. 73–79 and 81–87.

40. For the U.S. position, see, "Proclamation of President Nixon on Ratification of the Additional Protocol II to the Treaty for the Prohibition of Nuclear Weapons in Latin America," in *Arms Control and Disarmament Agreements: Texts and Histories of Negotiations* (Washington, D.C., U.S. Arms Control and Disarmament Agency, 1982) p. 78.

41. On this question, see Bertrand Goldschmidt, "France" in Jozef Goldblat, ed., *Non-Proliferation: The Why and the Wherefore* (see note 8 above) p. 69–70.

42. China's official policy is one of being critical about the NPT, but it professes not to be willing to assist others in acquiring nuclear weapons. This position has apparently been sufficient to enable the U.S. government to conclude an agreement for nuclear cooperation with the People's Republic. At a state dinner in Washington in January 1984, Premier Zhao made a toast, asserting, "We do not advocate or encourage nuclear proliferation. We do not do it ourselves, nor do we help other countries do it." And in an inaugural address to the General Conference of the IAEA in September 1984, the head of the Chinese delegation asserted that China would "take a discreet and responsible attitude so as to ensure that cooperation is solely for peaceful purposes," and that in exporting materials or equipment, China would request recipient states to accept IAEA safeguards. The Chinese position is discussed by Wu Xiu Quan, "China," and Reinhard Drifte, "China" in Jozef Goldblat, ibid., pp. 43–44; 45–56.

43. See Mohamed Ibrahim Shaker, *The Nuclear Non-Proliferation Treaty: Origin and Implementation 1959–1979* (London and New York, Oceana Publications, 1980).

44. Treaty on the Non-Proliferation of Nuclear Weapons, article X.2. The language of the article does not say the purpose is to decide *whether* to continue the treaty in force, but "whether the Treaty shall continue in force indefinitely, or shall be extended for an additional fixed period or periods," thus establishing a presumption of continuation.

45. If the NPT were to terminate, so would all of the safeguards agreements negotiated pursuant to the treaty. Unless fallback provisions existed (as would be the case with respect to all bilateral agreements that were suspended when NPT safeguards came into effect), a void would exist, creating very serious problems with respect to verification of use and material accountability.

46. U.S. Arms Control and Disarmament Agency, *Documents on Disarmament* (1969) p. 504.

47. This question is dealt with in chapter 6 of this book.

48. See Eisenhower Atoms for Peace speech (see note 32 above).

49. On the development and early years, see chapter 2 of this book; Bernard G. Bechoefer, "Negotiating the Statute of the International Atomic Energy Agency," *International Organization* (Winter 1959); Lawrence Scheinman, "IAEA: Atomic Condominium?" in Robert W. Cox and Harold K. Jacobson, eds., *Anatomy of Influence: Decision Making in International Organization* (New Haven, Conn., Yale University Press, 1972); and Paul Szasz, *The Law and Practices of the International Atomic Energy Agency* (Vienna, IAEA, 1970).

50. Elements of the latter can be found in United States General Accounting Office, "International Responses to Nuclear Power Reactor Safety Concerns," GAO/NSIAD-85-128 (Washington, D.C. GAO, September 30, 1985) and in the public positions adopted by groups such as the Sierra Club, National Resources Defense Council, and Friends of the Earth. Arguments to the effect that the IAEA could do more are normally to be found in statements of developing countries in the annual General Conferences of the IAEA. See, for example, statement of the representative of Malaysia in IAEA, GC(XXV)/OR.233 (1981), and of Jordan in IAEA, GC(XXVI)/OR.244 (1982).

51. IAEA, INFCIRC/254.

52. Article III.2.B of the Treaty on the Non-Proliferation of Nuclear Weapons is the relevant provision for the Zangger Committee.

53. See Leonard S. Spector, *Nuclear Proliferation Today* (New York, Vintage Books, 1984) pp. 446–447; and William C. Potter, *Nuclear Power and Nonproliferation: An Interdisciplinary Perspective* (Cambridge, Mass., Oelgeschlager, Gunn and Hain, 1982) pp. 44–45.

54. See IAEA, Regulations for the Safe Transport of Radioactive Material, 1985 edition, Safety Series No. 6 (Vienna, IAEA, 1985), for the question of transportation safety. Physical protection has been made the subject of an international convention that was opened for signature in March 1980 but still has not received the requisite number of ratifications (21) to come into force. Forty-five states and one international organization (EURATOM) have signed it, but thus far only 18 ratifications have been deposited. When in force, this convention will represent an important addition to nuclear safety and will also make a contribution to efforts to deal with the possible threat of nuclear terrorism.

55. The committee in question was established by resolution of the board of governors on April 1, 1970. It was open to participation by any member state desiring to be represented.

56. Statute of the International Atomic Energy Agency, article XI.

57. See, for example, the statement by the director general at the twenty-seventh General Conference, IAEA, GC(XXVII)/OR.247, paragraphs 81–84, October 10, 1983.

58. Environment and Natural Resources Policy Division of the Congressional Research Service, Library of Congress, *Nuclear Proliferation Factbook*, prepared for the U.S. Senate Subcommittee on Energy, Nuclear Proliferation and Federal Services of the Committee on Governmental Affairs, and the U.S. House Subcommittee on International Economic Policy and Trade of the Committee on Foreign Affairs (Washington, D.C., September 1983) p. 25ff and p. 32.

59. The words "it supplies" are italicized because while this offers a statutory authority upon which to build a repository if the member states so decide, the language was aimed primarily at material produced as part of an agency project rather than material that might for one reason or another be placed under agency safeguards by third parties.

60. For a discussion of these events, see Arnold Kramish, *The Peaceful Atom in Foreign Policy* (New York, Harper and Row, 1963) chapter 13.

61. Treaty on the Non-Proliferation of Nuclear Weapons, article III.2. For a comprehensive analysis of safeguards under the NPT, as applied by the IAEA, consult Paul Szasz, "International Atomic Energy Safeguards," in Mason Willrich, ed., *International Safeguards and Nuclear Industry* (Baltimore, Md., The Johns Hopkins University Press, 1973) pp. 73–142.

62. Statute of the International Atomic Energy Agency, article III.A.5.

63. A total of 54 project agreements were established with twenty-one member states over the twenty-five years between 1957 and 1982. Most of them involved small quantities of fissionable material for research purposes. See IAEA, *Review of the Agency's Activities* GC(XVIII)/718 July 1984, para. 152.

64. This system goes under the designation INFCIRC/66/REV.2, reflecting the extensions in 1966 and 1969 from reactors to reprocessing and fuel-fabrication facilities. It still is applied in agreements with states that have nuclear programs that have not submitted all of their nuclear activities to IAEA safeguards.

65. See U.S. Congress, House Committee on Foreign Affairs, Subcommittee on International Security and Scientific Affairs. Hearings on the Israeli attack on Iraqi nuclear facilities, June 17 and 25, 1981. U.S. Congress, Senate Committee on Foreign Relations. Hearings on the Israeli air strike, June 18, 19, and 25, 1981.

Chapter 2

FROM NEW YORK TO VIENNA: THE GENESIS OF THE IAEA

The bombing of Hiroshima and Nagasaki during World War II signalled the beginning of the era of atomic weaponry and dramatically revealed its unprecedented destructive potential. It was recognized that no one nation could claim a monopoly on nuclear weapons; nuclear science and technology was, or inevitably would be, dispersed around the world. And the vast potential of nuclear energy for military as well as peaceful applications appeared to provide irresistible incentives for its widespread development. On the basis of these perspectives, the international control of nuclear energy was advocated in the Agreed Declaration of Canada, the United Kingdom, and the United States; in the Moscow Declaration of the United Kingdom, the United States, and the Soviet Union; and in the U.S. Baruch Plan.

INTERNATIONAL CONTROL: THE FAILURE OF THE BARUCH PLAN

Early Allied Declarations

In the Agreed Declaration of November 15, 1945, U.S. President Harry Truman, British Prime Minister Clement Attlee, and Canadian Prime Minister Mackenzie King stated that they had conferred to consider "international action":

- To prevent the use of atomic energy for destructive purposes; and

- To promote the use of recent and future advances in scientific knowledge, particularly in the utilization of atomic energy, for peaceful and humanitarian ends.[1]

As noted in the preceding chapter, the parties recognized that "[n]o system of safeguards can be devised [that] will of itself provide an effective guarantee against production of atomic weapons by a nation bent on aggression."[2] Nevertheless, they expressed their willingness to exchange scientists and scientific information involving "practical industrial application of atomic energy with any reciprocating member of the United Nations just as soon as effective enforceable safeguards against its use for destructive purposes can be devised."[3] They also called for the United Nations to establish a commission to make specific proposals:

- For extending between all nations the exchange of basic scientific information for peaceful ends;
- For control of atomic energy to the extent necessary to ensure its use only for peaceful purposes;
- For the elimination from national armaments of atomic weapons and of all other major weapons adaptable to mass destruction; and
- For effective safeguards by way of inspection and other means to protect complying states against the hazards of violations and evasion.[4]

The Moscow Declaration added the Soviet Union's voice to the call for the creation of a commission on atomic energy under the auspices of the United Nations. At the Moscow Foreign Minister's Meeting in December 1945, the United States and the Soviet Union, along with the United Kingdom, forged a draft resolution on the establishment of a commission for the control of atomic energy, repeating the four areas of specific concern; the resolution gave rise to specific proposals that had been in the earlier Agreed Declaration of the three Western allies.[5]

The United Nations Atomic Energy Commission (UNAEC) was duly established by General Assembly Resolution 1, on January 24, 1946. Its origins were in the draft resolution agreed to in Moscow, which was later cosponsored by France and China, so it was not surprising that the UN General Assembly Resolution's terms of reference were identical with those specified in the Moscow and Three Nation Agreed Declarations. The UNAEC became the forum for the presentation of a bold U.S. initiative for the international control of atomic energy—the Baruch Plan—which was based on the Acheson-Lilienthal Report, an extensive and searching critical review of atomic energy issues.

The Acheson-Lilienthal Report and the Genesis of the Baruch Plan

In the expectation that the United Nations would act favorably on the proposal to establish the UNAEC, Secretary of State James F. Byrnes on January 7, 1946, directed Undersecretary of State Dean Acheson to chair a committee to formulate U.S. policy on international control of atomic energy. Acheson's committee, which included Vannevar Bush, James B. Conant, Gen. Leslie R. Groves, and John J. McCloy, named a Board of Consultants. Headed by David E. Lilienthal, the prestigious board members were J. Robert Oppenheimer, Charles Bernard, Charles Thomas, and Harry Winne. The result of the committee's frenetic work, known as the Acheson-Lilienthal Report, was completed in mid-March and released on March 28, 1946.[6] The report, in essence, proposed an international authority that would monopolize (own and manage, and not merely inspect) all dangerous atomic activities, while leaving safe and productive activities open to individual countries and private interests. As we have seen, the report made clear the inadequacies of inspection (that is, safeguards) alone, while recognizing that inspections would have to be a vital component of any system of international control.

From the beginning, it had been Acheson's intention that the committee produce a study of the controls and safeguards necessary to protect the United States government; he expected that the study would help the U.S. representative to the UNAEC. In March 1946, Bernard Baruch was appointed to that position. Although Baruch and his staff would propose substantive modifications to the Acheson-Lilienthal Report, Baruch ultimately supported the report's basic concept of an international agency with broad responsibilities for ownership and management, as well as research and development. The Baruch Plan, reflecting one of the report's conclusions—that the ability to produce special nuclear material was a critical step toward proliferation as, if not more, important than weapons design itself—called for the establishment of an international agency at the cutting edge of nuclear knowledge and with exclusive authority to conduct all dangerous (that is, militarily sensitive) operations in the nuclear field.

Presenting the Baruch Plan

In presenting the plan that would bear his personal imprint and his name before the UNAEC on June 14, 1946, Baruch dramatically depicted the dangers of the atomic age and declared that the world faced a choice between the "quick" and the "dead." To assure that atomic energy is used for peaceful rather than military purposes, he stated that the United States was proposing an International Atomic Energy Development Authority, which was to be entrusted with "all phases of the development and use of atomic energy."[7] Specifically, the authority's powers were to begin with the raw material itself and include:

- Managerial control or ownership of all atomic-energy activities potentially dangerous to world security.
- Power to control, inspect, and license all other atomic activities.
- The duty of fostering the beneficial uses of atomic energy.
- Research and development responsibilities of an affirmative character intended to put the authority in the forefront of atomic knowledge and thus to enable it to comprehend, and therefore to detect, misuse of atomic energy. To be effective, the authority must itself be the world's leader in the field of atomic knowledge and development and thus supplement its legal authority with the great power inherent in possession of leadership in knowledge.[8]

Baruch made clear that before the United States would halt manufacture of atomic weapons and dispose of its existing stockpile, and before the authority would be given full information, the world would have to agree on and put into effective operation "an adequate system for control of atomic energy, including the renunciation of the bomb as a weapon," and establish "condign punishments . . . for violations of the rules of control which are to be stigmatized as international crimes."[9]

The plan he outlined, Baruch said, was offered "as a basis for beginning our discussion."[10] But, it was not apparent at that time that the basic precepts of the plan were really negotiable. Although the Baruch Plan held out the hope for the prospect of beneficial peaceful uses of atomic energy, in its tone and substance it was primarily directed toward controlling the military applications of atomic energy and preventing other nations from possessing these unprecedented weapons of destruction. In the end, it was this sweeping objective, and the comprehensive control system it required for its realization, that would make the plan unacceptable to the Soviet Union.

The Soviet Reaction

Within days of the presentation of the U.S. proposal to the UNAEC, Andrei Gromyko, then the Soviet representative to the commission, also addressed that forum. Although he did not directly criticize the Baruch Plan, he offered instead a draft convention, the object of which would be "the prohibition of the production and employment of atomic weapons, the destruction of existing stocks of atomic weapons and the condemnation of all activities undertaken in violation of this convention."[11] Only later, according to Gromyko, should such an action be followed by "other measures aiming at the establishment of methods to ensure the strict observance of the terms and obligations in the . . . [proposed] convention, the establishment of a system of control over the observance of the convention and the taking of decisions regarding the sanctions to be applied against the unlawful use of atomic energy."[12] Gromyko's speech propounded proposals as unacceptable to the United States as Baruch's proposals had been to the Soviet Union. In doing so, this speech presented a clear indication of the major lines of Soviet criticism of the U.S. proposal that would emerge during its subsequent consideration by the UNAEC. In essence, the Soviets were unwilling to accept U.S. insistence on international inspection and control of atomic energy to prevent production of atomic weapons, holding that a declaration outlawing these weapons was sufficient. The Soviets were also adamantly opposed to the U.S. desire to give the international authority the right to impose sanctions, and its intention to permit no recourse to a veto in the Security Council.

Impasse in the UNAEC

Despite Soviet opposition and obstruction, the Baruch Plan was discussed for several months in the UNAEC, and reported on favorably (by a vote of ten "ayes," with the Soviet Union and Poland abstaining) in the UNAEC's first report to the United Nations Security Council.[13] The favorable findings of the first report notwithstanding, hope for an international control agency was fading by the end of 1946. It is by no means clear that the United States itself would have accepted an international regime that would have eventually transferred to an international authority the greater part of the U.S. atomic weapons complex (other parts having been dismantled, presumably), and probably would have required the reinstitution of a large standing army to defend U.S. interests in Europe and the Far East. But whatever action a future president or Congress might have taken, the current administration was in principle committed to such a regime, and it was primarily because of the fundamental and unalterable opposition of the Soviet Union that U.S. proposals for international control had little hope of realization in the early years of the Cold War. In effect, the U.S. position was that the Soviets "filibustered" negotiation on the U.S proposal in the UNAEC, especially through their introduction of essentially obstructionist counterproposals.[14] The Soviets cast the only negative vote against the second report of the UNAEC, which was completed on September 11, 1947.

Clearly, by 1948, policymakers in the United States had decided that there was no further basis for negotiations on international control of atomic energy in the UNAEC, primarily because the Soviet Union opposed essential elements of a system of control as supported by the Western and Third World states, and refused to accept the level of participation in the world community required in the field of atomic energy.[15] According to the U.S. government, fundamental points of disagreement between the majority (the United States and its allies) and minority (USSR) positions were those that emerged in the immediate aftermath of the presentation of the Baruch Plan, and included ownership of source material; ownership, management, and operation of "dangerous facilities"; research; elimination of atomic weapons from national armaments; inspection (safeguards); and enforcement (sanctions).[16]

The Final Report of the UNAEC

The U.S. government recognized an impasse in the UNAEC negotiations and recommended they be suspended, placing the blame for this situation on the intransigence of the Soviet Union. This position was largely reflected in the third and final report of the UNAEC, which stated:

> . . . [In] the field of atomic energy, the majority of the Commission has been unable to secure the agreement of the Soviet Union to even those elements of effective control considered essential from the technical point of view, let alone their acceptance of the nature and extent of participation in the world community required of all nations in this field by the first and second reports of the Atomic Energy Commission. As a result, the Commission has been forced to recognize that agreement on effective measures for the control of atomic energy is itself dependent on cooperation in broader fields of policy.

> The failure to achieve agreement on the international control of atomic energy arises from a situation that is beyond the competence of this Commission. In this situation, the Commission concludes that no useful purpose can be served by carrying on negotiations at the Commission level.
>
> The Atomic Energy Commission, therefore recommends that, until such time as the General Assembly finds that this situation no longer exists, or until such time as the sponsors of the General Assembly resolution of 24 January 1946, who are the permanent members of the Atomic Energy Commission, find, through prior consultation, that there exists a basis for agreement on the international control of atomic energy, negotiations in the Atomic Energy Commission be suspended.[17]

The report recognized that the scientific and technical requirements for the international control of atomic energy posed a formidable challenge to traditional precepts and prerogatives of national security and sovereignty, and to existing international political and economic practices. But, it argued on these grounds for effective international cooperation and the full exchange of information as a "novel approach" to "controlling a force so readily adaptable to warfare" and preventing "national rivalries in the most dangerous field."[18] The report elaborated on the pragmatic and idealistic motives behind effective international control in a manner that pointed to the later initiative of President Eisenhower known as Atoms for Peace:

> . . . all Members of the United Nations share the conviction that, unless effective international control is established, there can be no lasting security against atomic weapons for any nation, whatever its size, location, or power.
>
> . . . [A]tomic energy must not be developed on the basis of national interests and needs, means and resources, but that its planning and operation be made a common enterprise in all its phases.
>
> Only if traditional economic and political practices are adapted to the overriding requirements of international security, can these proposals be implemented. Traditional conceptions of the economic exploitation of the resources of nature for private or national advantage would then be replaced in this field by a new pattern of cooperation in international relations.
>
> Furthermore, secrecy in the field of atomic energy is not compatible with lasting international security. Cooperative development and complete dissemination of information alone promise to remove fears and suspicion that nations are conducting secret activities. . . . [T]he scientific and technical evidence makes such conclusions inescapable . . . .
>
> The new pattern of international cooperation and the new standards of openness in the dealings of one country with another that are indispensable in the field of atomic energy might, in practice, pave the way

for international cooperation in broader fields, for the control of other weapons of mass destruction, and even for the elimination of war itself as an instrument of national policy.[19]

The Atomic Energy Act of 1946 and the U.S. Policy of Secrecy and Denial

Even before the Baruch Plan died within the United Nations, the United States effectively ended its wartime atomic cooperation with its Western allies. Domestically, through the Atomic Energy Act of 1946,[20] Congress established a regime for atomic secrecy and denial as a means of preventing the spread of nuclear technologies and materials. The 1946 act prohibited any peaceful nuclear cooperation until Congress was satisfied that effective international safeguards were in place. The original strictures of the act hindered advantageous nuclear cooperation with our allies, especially Canada and the United Kingdom. Early on, pressures to amend the act resulted from the U.S. desire to provide nuclear weapons information to certain allies and to engage in mutually beneficial exchanges relevant to the production of uranium and plutonium to meet the requirements of the U.S. atomic defense complex, as well as to the development of fuel elements for submarine propulsion systems.[21] In 1951, these pressures resulted in an amendment to the act to allow limited nuclear cooperation under agreements for cooperation.[22]

INTERNATIONAL COOPERATION: ATOMS FOR PEACE AND THE CREATION OF THE INTERNATIONAL ATOMIC ENERGY AGENCY

This relaxation of restrictive statutory provisions on nuclear cooperation and exports revealed serious problems with the policy of secrecy and denial and suggested that it would soon founder. Indeed, the shape of things to come—the shift in U.S. nuclear policy from denial to safeguarded nuclear assistance and cooperation—was already foreshadowed. During the period that followed the failure of the Baruch Plan, the Soviet Union and then the United Kingdom successfully tested nuclear explosive devices while several other countries (including Belgium, Canada, France, and Italy) established and set in motion national nuclear programs. These events demonstrated the obvious ineffectiveness of a policy of secrecy and denial, as institutionalized in the 1946 act, to prevent the spread of nuclear information and technology, let alone the development of nuclear weapons.

And they revealed growing interest and activity abroad in developing the commercial possibilities of this new technology, which could not be fully exploited under existing U.S. law and policy. At this time, it will be recalled, nuclear energy was *the* high technology, and there were fears that one country would develop its commercial uses and have a commanding advantage. In particular, the United States was concerned that its British allies might capture the anticipated world market with their gas-cooled reactors. As a consequence, the pressures for change in U.S. policy increased and ultimately prevailed in 1954.

Eisenhower's Atoms for Peace Speech

The new nuclear policy was in essence articulated in the important U.S. initiative embodied in President Dwight D. Eisenhower's Atoms for Peace address to the United Nations General Assembly on December 8, 1953. In his words:

> I make the following proposals:
>
> The governments principally involved, to the extent permitted by elementary prudence, to begin now and continue to make joint contributions from their stockpiles of normal uranium and fissionable materials to an International Atomic Energy Agency. We would expect that such an agency would be set up under the aegis of the United Nations. . . .
>
> The United States is prepared to undertake these explorations in good faith. Any partner of the United States acting in the same good faith will find the United States a not unreasonable or ungenerous associate.
>
> Undoubtedly initial and early contributions to this plan would be small in quantity. However, the proposal has the great virtue that it can be undertaken without the irritations and mutual suspicions incident to any attempt to set up a completely acceptable system of worldwide inspection and control.
>
> The Atomic Energy Agency could be made responsible for the impounding, storage, and protection of the contributed fissionable and other materials. The ingenuity of our scientists will provide special safe conditions under which such a bank of fissionable materials can be made essentially immune to surprise seizure.
>
> The more important responsibility of this Atomic Energy Agency would be to devise methods whereby this fissionable material would be allocated to serve the peaceful pursuits of mankind. . . . A special purpose would be to provide abundant electrical energy in the power-starved areas of the world. Thus the contributing powers would be dedicating some of their strength to serve the needs rather than the fears of mankind.

> The United States would be more than willing—it would be proud to take up with others principally involved in the development of plans whereby such peaceful use of atomic energy would be expedited.[23]

Eisenhower's proposal was conceived, developed, and presented to the United Nations at a time of urgent and deep concern about the nuclear weapons peril. He spoke frankly of the "atomic danger" and publicly acknowledged what the entry of the United Kingdom and the Soviet Union into the "Nuclear Club" had already made apparent to everyone that "the dread secret, and the fearful engines of atomic might" were no longer a U.S. monopoly.[24] The "atomic realities" of the day revealed two fundamental facts:

> First, the knowledge now possessed by several nations will eventually be shared by others—possibly all others.
>
> Second, even a vast superiority in numbers of weapons, and consequent capability to devastating retaliation, is no preventive, of itself, against the fearful material damage and toll of human lives that would be inflicted by surprise aggression[25]

Although facing the atomic danger, Eisenhower did not wish to succumb to its horrors. And, however frightful the reality, he refused "to confirm the hopeless finality of a belief that two atomic colossi are doomed malevolently to eye each other across a trembling world."[26] He sought not merely the "reduction or elimination of atomic materials for military purposes."[27] He stated: "It is not enough to take this weapon out of the hands of soldiers. It must be put into the hands of those who will know how to strip its military casing and adapt it to the arts of peace."[28] It was as a step toward this end that the proposal embodied in the president's speech was directed; and, perhaps to assure that this first step be taken, the stringent safeguards requirements of the Baruch Plan were essentially abandoned. The speech did allude to safeguards and security measures when President Eisenhower stated that the agency he proposed "could be made responsible for the impounding, storage, and protection of the contributed fissionable and other materials," and spoke of developing "special conditions under which such a bank of fissionable material can be made essentially immune to surprise seizure."[29] But the objective of creating a comprehensive international safeguards system was deemed impractical and counterproductive at that point. Whatever Eisenhower intended with respect to safeguards in his speech, the proposal developed into a program where safeguards would become one of its essential elements, albeit rather different from what had first been proposed in the Acheson-Lilienthal Report.[30]

Reactions to the President's Proposal

With its high ideals and its message of hope, peace, and prosperity, the Atoms for Peace address was received enthusiastically by the world. As former Atomic Energy Commission (AEC) Chairman and Secretary of Energy James Schlesinger remembered the world's response: "Today one can scarcely recall the drama of the occasion, the momentous impact of his comments, the enthusiasm with which they were greeted, the widespread belief that we were witnessing a Marshall Plan for atomic energy."[31] In the same vein, Richard Hewlett, who was the chief historian of the AEC and its successors, the Energy Research and Development Administration (ERDA) and the Department of Energy (DOE), wrote: "The worldwide outpouring of positive reaction to the speech demonstrated that the President had struck a highly resonant note at precisely the right moment."[32]

Although the Soviet Union would also praise the president's vision—it was impossible politically to impugn his idealism—it would appear that the initial Soviet response to the address was an attempt to use his proposal to obtain political and propaganda advantages. While the Soviet statement aligned itself with the ideals presented in the president's speech and expressed a willingness to participate in discussions on his proposal, it questioned whether the means proposed in President Eisenhower's address would truly realize the ideals he presented. Rhetorically questioning the meaning of the proposal, the Soviet statement asserted that:

> . . . [I]t means that it is proposed that only "some" small part of the existing stockpiles of atomic materials and of those to be created should be allocated for peaceful purposes. It follows from this that the bulk of atomic materials will continue to be directed to the production of new atomic and hydrogen bombs, and that the full possibility remains for the further accumulation of atomic weapons and the creation of new types of this weapon of still greater destructive power. Consequently, in its present form this proposal in no way ties the hands of the States which are in a position to produce atomic and hydrogen weapons.
>
> Secondly, President Eisenhower's proposal in no way restricts the possibility of using the atomic weapon itself. Acceptance of this proposal in no way restricts an aggressor in the use of the atomic weapon for any purpose or at any time. Consequently this proposal in no way reduces the danger of atomic attack.[33]

On the basis of this assessment, the Soviet Union drew the conclusion that the Atoms for Peace proposal, as presented in President Eisenhower's speech, "neither halts the growing production of atomic weapons nor restricts the possibility of their use. This must not be overlooked in making an appraisal of the real significance of this proposal."[34]

The Soviets then reiterated their proposals for the prohibition of atomic and hydrogen weapons, proposals that have their origin in the days of the Baruch Plan, and which would appear again and again in the next years. But the Soviet position, as presented in December 1953, sought to use the popularity of Eisenhower's proposal to push their own disarmament program. Thus, the Soviets stated that:

> Since we are striving to strengthen peace, our tasks cannot permit either the relaxation of vigilance in relation to the danger of atomic war or international sanction of the production of atomic weapons. It is precisely for this reason that it must be recognized that the task of all peace-loving States is not restricted to the allocation for peaceful purposes of a certain small part of atomic materials. Not a certain part, but the whole mass of atomic materials must be directed in its entirety to peaceful purposes, and this would open up unprecedented possibilities for developing industry, agriculture and transport, for applying invaluable atomic discoveries to medicine, for improving technology in many fields of application, and promoting further and greater scientific progress.
>
> It should also be borne in mind that the prohibition of atomic and hydrogen weapons and the use of all atomic materials for peaceful needs of the peoples, displaying due concern for the needs of the economically more backward areas, would at the same time improve the possibility of agreement on the question of a decisive reduction in conventional armaments.[35]

The president's proposal offered a fresh start for dealing with the conundrum of problems that had emerged out of the harnessing of the atom. Irrespective of the reactions of the Soviet Union and other states (favorable or not), and of the fact that its implementation would require at the outset a change in national law and the creation of an international organization (formidable tasks on the face of it), Eisenhower's plan was more practicable and negotiable than the Baruch Plan, if also much more modest in scope and import. The president's personal perspective on (and preoccupation with) the nuclear dangers posed by possession of atomic weapons, as well as his recognition that it was time to find ways to exploit nuclear technology for peaceful purposes (without jeopardizing a nation's security interests) led to his momentous (in spirit), albeit modest (in substance), proposal.

An Alternative Approach

The Atoms for Peace program could trace its lineage to "Operation Candor" and other Eisenhower initiatives to deal with the problem of nuclear weapons.[36] However, it was by no means the necessary harvest of a historical sowing. There had been other alternatives to the failed U.S. nuclear policy of the early postwar period. In the early 1950s, the idea of international control of nuclear energy had not been abandoned by everyone. W. Sterling Cole reports being approached by Edward Teller in the fall of 1953, when he was chairman of the Joint Congressional Committee on Atomic Energy. Teller was upset about the failure of the United States to reach agreement on international control of nuclear energy as presented in the Acheson-Lilienthal Report. According to Cole's account, Teller insisted that the only opposition to the approach advocated in the Acheson-Lilienthal Report was from the Soviets, and argued that the rest of the world should decide to pursue this path. If the Soviets desire to remain adamant, Teller declared, then the rest of the world should not be affected by their action. But, he said, world opinion might eventually pressure the Soviets to participate in an international system of effective control. Cole was impressed with Teller's reasoning and the appeal of the course of action he advocated, and he decided at that time to consider making a public statement recommending their serious assessment. He visited the secretary of state and the chairman of the AEC. AEC Chairman Lewis Strauss listened and indicated approval, but requested that Cole say or do nothing, because Strauss had to go to Bermuda to attend a meeting between President Eisenhower and the British prime minister. It was at the time of the Bermuda Conference that Eisenhower was finalizing his Atoms for Peace address, although Cole was not then aware of the president's intentions.[37] Of course, Atoms for Peace was the alternative policy developed and decided on by Eisenhower, and we must recognize that absent Eisenhower's personal initiative, it might never have been put forward.

Objectives of Atoms for Peace

Atoms for Peace was not merely an acknowledgment that we would be unable to develop a comprehensive international control regime. Nor was it simply a U.S. acquiescence in global trends of technology dissemination, although the gradual and visible erosion of American technological monopoly played a significant role in the scope and character of arrangements for international cooperation developed on the basis of the president's decision.

Eisenhower's motives and goals were complex. Derived from idealism and pragmatism, and made possible by America's unique position in the postwar world, Eisenhower's objectives also included the desire to strengthen and expand U.S. military and economic ties around the world, to assure American primacy in international nuclear councils, and to promote American power reactor sales.[38]

As we have suggested, however, Atoms for Peace, as it was originally conceived in the aftermath of the Soviet thermonuclear test, was aimed at moving beyond the impasse that characterized the Soviet-American nuclear dialogue. According to Jack Holl, chief historian of the DOE, "Eisenhower was determined to develop a peaceful alternative to nuclear weapons, and at the same time to use Atoms for Peace to break the deadlock in disarmament negotiations with the Soviet Union."[39] Holl's predecessor, Richard Hewlett, also has argued that the president's primary motivation in putting forward the Atoms for Peace concept was to promote disarmament and reduce the risks of nuclear war. Of this he stated:

> For Eisenhower, nothing was more important than reducing the threat of nuclear war. He never ceased to probe new possibilities for an agreement with the Soviet Union. He badgered his cabinet and the National Security Council to come up with new approaches to the problem. Although he was never willing to sacrifice American security or allow himself to be hoodwinked by the Russians, he was more often than most of his advisors prepared to take small risks that might lead to a test ban or put the world on the road to disarmament. He patiently endured the political and diplomatic gaffs of Harold Stassen because he could count on his disarmament advisor to prod other members of his official family to consider ways of resolving the nuclear stalemate.[40]

The Atomic Energy Act of 1954

In light of the circumstances of its origin and presentation, no less than its primary objective, it is not surprising that the president's proposal had limited domestic goals. But, as suggested, it did require a revision of the Atomic Energy Act of 1946. On February 17, 1954, Eisenhower proposed to Congress a major revision of the 1946 act. In his message to Congress, the president argued that his recommended amendments to the Atomic Energy Act of 1946 were designed to strengthen the defense and economy of the United States, its allies, and friendly states, and with proper safeguards, would accomplish this objective by:

- First, widened cooperation with our allies in certain atomic energy matters;
- Second, improved procedures for the control and dissemination of atomic energy information; and,
- Third, encouragement of broadened participation in the development of peacetime uses of atomic energy in the United States.[41]

Hearings were held by the Joint Committee on Atomic Energy in May and June, during which some concern was expressed about a possible "giveaway" of U.S. atomic secrets, nuclear materials, and technologies. With Chairman Cole effectively countering these arguments, and administration witnesses touting the benefits of the proposed changes for world peace and the strength and prosperity of the United States and all of the free world, the legislation was passed by Congress and, on August 30, 1954, was signed into law.

Although Eisenhower's legislative initiative did not mention the international agency he had proposed in his UN speech, it did effectively clear away obstacles to the eventual creation of the agency.[42] The Atomic Energy Act of 1954 opened the way to expanded international cooperation, and thereby facilitated the Atoms for Peace objective of peaceful nuclear cooperation under safeguards. Most notably, it authorized the AEC to cooperate with other nations to further peaceful uses of nuclear energy to the extent that expanding technology and the consideration of the common defense and security would allow. Such collaboration would take place under agreements of cooperation with individual nations or groups of nations. The act required that all such agreements include a pledge by the cooperating party that any materials or equipment supplied by the United States would be used only for peaceful purposes and that safeguards as provided in the agreement would be maintained.

Implementing Atoms for Peace: Toward the Creation of the IAEA

Although it had significant domestic repercussions, the president's proposal was primarily directed outwards, as has been indicated. Its foreign or international objectives and, in particular, the disarmament objective, were to deeply affect the style, scope, and substance of the negotiations leading to the creation of the institutional apotheosis of Atoms for Peace—the International Atomic Energy Agency (IAEA). As Bernard G. Bechhoefer, who was a special assistant to the U.S. representative to those negotiations, would later observe:

> If the agency were to serve no purposes except to promote the peaceful uses of atomic energy with particular emphasis on the underdeveloped areas of the world, it would have been desirable to negotiate rapidly with a minimum of negotiating machinery. The sooner that an agency was organized and in a position to operate, the more rapid the progress could be. On the other hand, if, at the same time, one objective was to create better understanding between the east and the west and ultimately to reduce the menace of nuclear warfare, then at all stages of the negotiations, there should be consultation and close understanding with the Soviet Union. Anyone familiar with the Soviet Union's negotiating techniques will realize that this could be accomplished only at the expense of speed and efficient negotiations. Furthermore, this objective tended to shift the emphasis from the positive aspect of developing peaceful uses to the negative aspect of "safeguards" to prevent diversion for military purposes.[43]

In his Atoms for Peace address, Eisenhower implied that the United States would be willing to engage in private discussions with the states "principally involved, " especially the Soviet Union, to develop plans to expedite the objectives he put forward. And, as indicated, the Soviets expressed a willingness to engage in such conversations. But it was first necessary to delineate the U.S. position.

A Draft U.S. Statute for the Agency

Clearly, with or without the Soviets, carrying out the Atoms for Peace proposal as originally developed and delivered by the president would require the establishment of an international nuclear pool under the supervision of an international atomic energy agency. The fissionable material deposited with this agency would be devoted to peaceful purposes and, in theory, not available for use in weapons. Although it was clear that this agency and its "atomic pool" were to be the focus of Atoms for Peace, the president's address provided no other explicit directives for the implementation of the proposal. Consequently, as one analyst has argued, "in 1954 the AEC and the Department of State faced the difficult task of devising specific programs to accomplish the broad objectives set forth in the President's speech. Caught in this policy vacuum, the executive agencies faced the President's commitment with no clear understanding of his intentions and, indeed, without any determination on their part that his proposals were feasible."[44] It was this situation, according to the analyst, that serves at least in part to explain "why it took many months to formulate programs, and why the results were so long in coming."[45]

However long they required, these agencies did finally produce a draft statute for the agency. An outline of the proposed agency, written by the AEC and modified by the Department of State, appeared to embody the president's vision as expressed in the December speech. It proposed contributions of uranium and fissionable material to be allocated by the agency for peaceful purposes. Specifically, this agency was "to receive supplies of nuclear materials" to be used:

A. to encourage world-wide research and development of peaceful uses of atomic energy by assuring that engineers and scientists of the world have sufficient materials to conduct such activities and by fostering interchange of information;
B. to furnish nuclear materials to meet the needs of agriculture, medicine, and other peaceful activities including the eventual production of power.[46]

The most important responsibilities of the agency were delineated in Section III.B.3 of the outline. There, in addition to providing the agency with the authority to undertake such actions as prescribing "certain design and operating conditions" and "health and safety regulations," and specifying the "disposition of by-product fissionable materials and wastes," the outline provided for the agency's authority to "verify status of allocated material inventories and to verify compliance with the terms of issuance," to "require accountability and operating records," and to "retain the right of monitoring and require progress reports."[47] Clearly, there was a world of difference between those provisions of the outline that were designed to verify or assure that allocated fissionable material was being used for the purpose for which it was intended, and the comprehensive control over such material specified in the Baruch Plan. Although such "safeguards" would prove to be more negotiable than the control provisions of the Baruch Plan, the infringement on national sovereignty that they implied would by no means be immediately welcomed by the world community. This became evident when the outline was appended to a memorandum passed from Secretary of State John Foster Dulles to Soviet Ambassador Zaroubin on March 19, 1954.

U.S.-Soviet Discussions

Although exchanges between the United States and the Soviet Union on the international agency had occurred since January, the March 19 memorandum was the first to provide the Soviets with an idea of the proposed functions and responsibilities of the new agency, as well as such matters as its administration, financing, and relationship to the United Nations. As one participant in the process observed, this memorandum "purposely avoided rigid solutions at this early stage."[48] The Soviet response, which took the form of an aide-memoire handed to Secretary Dulles from Soviet Foreign Minister Vyacheslav Molotov on April 27, 1954, did not comment specifically on the U.S. proposals. Rather, the Soviets stated that the proposed agency would not halt the arms race or hasten disarmament, because the applications of atomic energy for peaceful purposes could be diverted for weapons. While the Soviet aide-memoire rightly recognized the dual nature of atomic energy, and pointed to this fact as a potentially serious problem with Atoms for Peace as a disarmament proposal, the USSR's intention was clear to the U.S. participants: they intended either to hold the agency negotiations hostage to, or to bypass them in favor of, their own vague disarmament proposals.[49] This conclusion was based on the considerable attention the Soviets gave to their own disarmament proposals in the aide-memoire, and their earlier insistence on considering their proposal for a great power pledge not to use atomic and hydrogen bombs or other weapons of mass destruction, into agency negotiations. This may be attributed, in part, to the provisions for safeguards contained in the U.S. document—provisions that, however vaguely defined, the Soviets found unacceptable.

The United States had already been deemphasizing the disarmament concept that appeared as the basic rationale of the Atoms for Peace speech, and highlighting those aspects of the speech having to do with peaceful nuclear cooperation. And, during the summer of 1954, the United States responded to these Soviet charges by denying that the proposal for the agency was in itself intended as a disarmament measure. In light of the Soviet position, the United States declared that with or without Soviet participation it might examine the establishment of the agency with other interested nations. A memorandum of July 9, 1954, by Assistant Secretary of State Livingston Merchant to Soviet Ambassador Zaroubin, clarified the U.S. position. According to Merchant, the Atoms for Peace proposal

> was intended to make a beginning toward bringing to the peoples of the world the peaceful benefits of atomic energy. This offer by the

> United States to join with other nations having atomic facilities to furnish fissionable material and atomic energy technology for the common benefit, would provide a new opportunity for international cooperation. Successful cooperation in the implementation of the President's proposal would surely result in an improved atmosphere, which, in turn, could significantly improve the prospects for genuine, safeguarded international disarmament. The proposal itself was not put forward as a disarmament plan.[50]

Merchant then proceeded to explain how the Soviet aide-memoire misconstrued that position, arguing that the Soviet document essentially stated that the USSR would not cooperate in measures to achieve the peaceful benefits of atomic power until the United States agreed to ban the use of atomic weapons. According to Merchant, the primary reason for the Soviet position, as presented in the aide-memoire, was that stockpiles of weapon-grade material could in principle increase even after the international agency envisioned in the president's speech was established. Yet, Merchant argued, the Soviet proposal would not prevent increases in stockpiles, and therefore did not provide a valid objection to proceeding with steps for promoting the peaceful uses of atomic energy. According to Merchant:

> The Soviet Union also appears to assume that any form of peaceful utilization of atomic energy must necessarily increase stocks of materials available for military purposes. In reality, however, ways can be devised to safeguard against diversion of materials from power producing reactors. And there are forms of peaceful utilization in which no question of weapon grade material arises.[51]

Reiterating the obligation of the United States and other advanced nuclear nations to make the benefits of peaceful uses of atomic energy available under appropriate conditions, Merchant concluded that "the United States will feel free to go ahead with its proposal with other interested nations."[52] Nonetheless, he stated that: "If at a later time the Soviet Union should decide to take part in any such discussions, the United States will continue to welcome such participation."[53]

On September 22, 1954, the Soviets presented the United States with an aide-memoire that constituted a *volte-face*. Whatever the reason for the change—and it appears to have been inspired, at least in part, by the desire to avoid the political damage that could be expected by their blocking a positive initiative that had received worldwide acclaim—the Soviets argued in essence that there were technical means that could prevent materials located at peaceful facilities and intended for peaceful purposes from being used for military ends, and that they were ready to examine the U.S. position on such safeguards in the context of further negotiations. This Soviet interest in technical aspects of safeguards is reminiscent of the June 11, 1947 Soviet counterproposal of the Baruch Plan, which called for an International Control Commission (ICC) with "its own inspectorial apparatus" and provided that "the ICC shall periodically carry out inspection of facilities for mining of atomic raw materials and for the production of atomic materials and atomic energy."[54] Whatever the import of such precedents, the Soviet shift allowed the convocation of a confidential meeting on safeguards in Geneva from August 22 to 27, 1955, with the United States, the Soviet Union, the United Kingdom, France, Canada, and Czechoslovakia in attendance. Participants discussed the safeguards that would be required for peaceful uses of atomic energy undertaken through an international agency. It is notable that the U.S. participant and possibly the participants from other countries concluded from the discussions that physical security complemented by material accounting would be required for the international agency's safeguards regime, a theoretical conclusion that would later be reflected in the practice of agency safeguards.[55]

Multilateral Negotiations

Although the private exchanges between the United States and the Soviet Union on the creation of an international atomic energy agency continued during 1954, these bilateral "negotiations" proceeded at a snail's pace. In May 1954, the United States already was moving to examine the creation of such an agency with other "interested nations" (advanced nuclear and uranium supplier states), all of which were from NATO and the British Commonwealth. Although active negotiations between the United States, the United Kingdom, France, Canada, Australia, Belgium, South Africa, and Portugal did not actually begin until late 1954, knowledge of the existence of this narrowly composed and unrepresentative negotiating group would give rise to serious objections during the Ninth Session of the General Assembly in the fall of 1954. Notwithstanding these objections, the General Assembly was enthusiastic about the prospect of an international agency. It expressed the hope that the agency would be established without delay, that UN members would be informed of progress and their perspectives taken into account and that, once established, the agency would negotiate appropriate agreements with the UN.[56] In the same vein, the General Assembly endorsed a proposed international scientific conference on peaceful applications of atomic energy, to be held in Geneva no later than August 1955, and established a Scientific Advisory Committee (SAC) reporting to the UN secretary general to plan this conference.

The negotiations among the group of eight negotiating states began in December 1954. A first draft of the statute was produced by the United States. Taking account of the suggestions of other negotiating states and of the UN General Assembly debates in its preparation, the draft was submitted to the eight negotiating states on March 29, 1955. During April and May, the United States discussed the draft with interested domestic and international parties. Suggested changes were agreed upon by the eight and the resulting draft statute was submitted on a confidential basis on July 29, 1955, to the Soviet Union, which had on the 18th of July, during the Geneva Summit, indicated its readiness to participate in negotiations on the agency. On August 22, 1955, the draft with minor amendments was submitted on a confidential basis to all eighty-four members of the United Nations or the specialized agencies, with a request for comments from all states.[57]

The Geneva Conference and UN Activity

It was at this time that the highly successful Geneva Conference on the Peaceful Uses of Atomic Energy was held. During this international scientific conference, newly declassified data on atomic energy was widely disseminated, dramatic advances in peaceful applications were described, and tremendous future benefits of the peaceful atom were touted.[58] In the euphoria and enthusiasm for the peaceful atom that followed the Geneva Conference, the Tenth Session of the UN General Assembly, held in the fall of 1955, could consider the peaceful uses of atomic energy and the proper procedures for establishing the agency under a propitious star. Participants in the Geneva Conference were congratulated for their successful and useful effort by the General Assembly, which endorsed the continuation of the SAC that had prepared the conference. In the First Committee, comments were heard on the August draft of the agency's statute, and it was concluded that an international conference would be held in September 1956 to debate the draft and agree on a final text for the statute.

The Twelve-Power Conference

In preparation for that conference, twelve nations agreed to hold a drafting session in Washington, D.C., early in 1956. The twelve-power working group, which included the original group of eight, joined by the USSR, Czechoslovakia, Brazil, and India, began their work on February 27, 1956, and concluded it on April 27 of that year. The draft they produced was substantially different from that produced by the group of eight in August 1955. Not only did it increase the membership of the board of governors and introduce a two-budget system (assessed contributions for administrative expenses and voluntary contributions for operational expenses), but it also clarified the safeguards provisions of the earlier draft. It was decided that the obligation of a state to submit to safeguards did not derive from membership in the agency, but from the application for and reception of agency assistance. In effect, this meant that the great powers—including the United States, the Soviet Union, and the United Kingdom, which were not expected to apply for assistance from the agency—would not be subject to agency safeguards.[59] The draft agreed to by the twelve was submitted to all UN member states and specialized agencies, and became the basis for debate at the international conference held in the fall.

The Conference on the IAEA Statute

The conference opened at UN headquarters on September 20, 1956, and was attended by eighty-two nations and seven specialized agencies. The UN secretary general served as the secretary general of the conference, and Ambassador João Carlos Muñiz of Brazil and Dr. Pavel Winkler of Czechoslovakia were elected as conference president and vice president, respectively, during the opening plenary sessions. Under their leadership, the conference proceeded rapidly and relatively smoothly toward its appointed objective.

The twelve negotiating powers, which served as sponsors of the conference, held that because of the compromises contained in the draft it should not be substantively revised during the coming conference, and actively opposed any revisions. There was a widespread recognition among the participants at the conference that the draft statute before them was indeed a compromise document, and a great many states concurred with the position of the twelve negotiating nations on changes to the document. States speaking out in favor of the draft during the early plenary sessions included Chile, Greece, Norway, and Australia, as well as the original negotiators (such as France, Portugal, South Africa, and the United Kingdom). This favorable predisposition of many of the participants was reflected in the rules of procedure drawn up by the coordination committee on

September 21, which required a two-thirds majority to modify the draft. As it was, more than eighty amendments to the draft would be considered by the conference, and about sixty would be passed. Nevertheless, most of the amendments were relatively uncontroversial, and the statute endorsed by the conference was basically that completed in April.

Attitudes and Issues[60]

The speeches delivered during plenary sessions in late September and early October revealed the expectations of the participants, and foreshadowed the scope and substance of the conference. The Hungarian revolution of 1956 did not create an East-West confrontation at the conference, but in the plenary speeches there was evidence of "politicization." Most notably, the Soviet Union and its satellites attempted to use the conference to reiterate their demand for a ban on atomic and hydrogen bombs and, supported by a host of Third World countries, regretted the absence of the People's Republic of China and the German Democratic Republic from the proceedings. Despite such political tensions, the necessary Soviet cooperation with the United States and its allies in the common pursuit—creating an international machinery that would prove able to serve the interests of the superpowers and all other states in the coming decades—was forthcoming.

With East-West differences reduced to acceptable levels in New York, the conference was able to recreate to a large extent the exhilaration produced by President Eisenhower's Atoms for Peace address. This, too, was evident in the opening speeches of the delegations, in which broad hopes were expressed that the peaceful pursuit of the atom would unify the world, raising it to a higher level technologically, economically, and politically. The United States and New Zealand expressed the hope that the steps taken by the agency might take the world down the path toward full disarmament. Hopes were also expressed by such less-developed states as Iran, India, and Guatemala that technical assistance from the IAEA would be channeled to encourage development in the Third World, and that the cost of assistance would be related to a state's ability to pay. But, foreshadowing later fears, Israel and Japan expressed some concern about assurances of supply, and about the prospect of being dependent upon the whims of a major supplier if assurances or guarantees were not provided.

There was widespread recognition of the need for safeguards—expressed by such states as Belgium, Norway, Spain, Japan, Thailand,

Pakistan, and Jordan—although there were also considerable differences over their application. Concern that safeguards might infringe on sovereign rights was expressed by such diverse countries as Iran, India, Japan, and Bulgaria. And, interestingly, the representatives of Ceylon (now Sri Lanka) and India argued that the safeguards provisions of the agency should be applicable to all states, and not merely to recipients of assistance, in order to avoid a situation in which there were two levels of inspections and control.

The discussions in the main committee over individual articles (which ran from October 3 to October 19) were characterized by good will and a constructive spirit. Although differences would be heard, only safeguards were highly controversial. There was a host of amendments to Article III dealing with the functions of the IAEA, but they were not substantive and were easily resolved. The discussion of amendments to Articles V, VI, and VII, which deal with the general conference, the board of governors, and the staff, respectively, revealed differences between the nuclear "haves" and "have-nots" over representation on the board of governors, the relationship between the board and the general conference (with many of those excluded from the former arguing for more powers for the latter), and the appointment and reporting responsibilities of the director general. However, the twelve-nation compromise held up, principally as a consequence of the assurances and arguments presented by one or another (or all) of the twelve.

As indicated, safeguards were the most controversial issue addressed at the conference. One subject of controversy was whether source materials (for example, uranium as it occurs in nature and other materials as determined by the board) should be safeguarded. The Belgians, Australians, Canadians, and British were among those who would have the agency's safeguards apply to source materials; the French, Indians, and others opposed this position. In the end, the effort led by France to change the safeguards provisions in the draft statute was unsuccessful.

Perhaps the more important controversy was over the nature of safeguards themselves. The Indians, supported by the Soviets, decried the proposed safeguards as an interference by the IAEA in the economic development of member states; and the United States and its principal allies, Canada and the United Kingdom, argued that the safeguards provided for in the draft statute were the minimum needed to assure that assistance provided by the IAEA would be used solely for peaceful purposes. Carefully crafted language reconciled these differences.[61] The safeguards and other outstanding issues (for instance, the composition of the board and its relationship to the general conference) having been satisfactorily resolved, the conference reconvened in plenary session. And, on October 26, 1954—after a message from President Eisenhower offering to place 5,000 kilograms of U-235 at the disposal of the agency, and promising to match all other contributions made to the agency, was read—the Statute of the IAEA was signed by seventy nations and the international conference concluded. As provided by Annex I of the statute, a preparatory commission, to be composed of the group of twelve and six other states chosen by the conference, was charged with developing a program and arranging the first general conference and the first meeting of the board of governors.

U.S. Ratification of the Statute

In the United States, President Eisenhower submitted the International Statute to Congress on March 21, 1957. During hearings before the Senate Foreign Relations Committee and Members of the Joint Committee on Atomic Energy, the administration engaged in a "hard sell" to counter some congressional concerns, arguing that approval of the statute would accelerate nuclear progress, provide safeguards, improve nuclear health and safety, strengthen control of nuclear weapons, reduce pressure for proliferation, and improve the climate of international relations.[62] In addition to indicating the benefits of approval, administration witnesses also warned of the negative consequences of a failure to ratify the statute. Not only would the notable benefits of the agency be sacrificed, they argued, but the loss to U.S. prestige and influence would be incalculable, and the Soviets would gain the initiative in atomic matters.[63] Finally, the administration witnesses again attempted to allay openly expressed concerns in some quarters of Congress that the agency constituted an atomic "give-away"; they asserted that U.S. secrets would not be compromised, that there were limitations on the U.S. commitment of support, that most contributions were voluntary and the United States would pay no more than its share of administrative costs, and that countries receiving materials from the agency would have to pay for them.[64] Although the chances of congressional non-ratification did not appear to be great, this set of administration arguments was apparently well chosen and convincing. On June 18, 1957, the Senate gave its advice and consent to ratification, and the next day a bill to provide for U.S. participation in the IAEA was introduced and, after hearings, was duly enacted.[65]

The Birth of the IAEA and the First General Conference

With the required eighteen ratifications (three of which had to come from the United States, the United Kingdom, the Soviet Union, France, or Canada), the agency came into being on July 29, 1957. The first general conference opened on October 1, 1957, in Vienna, the site recommended by the statute conference. Presided over by Karl Gruber, the Austrian ambassador to the United States, the general conference proceeded with no serious problems. Admiral Lewis Strauss, the chairman of the U.S. Atomic Energy Commission, formally presented the U.S. offer of fissionable material available to the agency. Sterling Cole, the U.S.-promoted candidate who had been appointed by the board of governors, was confirmed by the conference as the first director general of the IAEA. In informal discussions preceding the conference, several delegations met with Soviet Representative V.S. Emelyanov, and were apparently able to overcome initial Soviet objections to Cole's candidacy. The first meeting of the board of governors was held in January 1958.

Initial U.S. Assessments of the New Agency

The statute produced an organization that may not have fully reflected the vision of President Eisenhower's Atoms for Peace proposal. Notwithstanding, Eisenhower congratulated the conference for its accomplishments, stating that the "Statute, and the International Agency for which it provides, held out to the world a fresh hope of peace."[66] He stated that "people have long been seeking a channel for peaceful discussion," and declared that the IAEA "offers one such channel."[67] He argued that the "prompt and successful functioning of the Agency can begin to translate the myriad uses of atomic energy into better living; in our homes, at our work, during our travel and our rest. . . . We will not lead people to expect the atomic millennium. . . . But . . . this International Agency will be encouraging those scientific labors and research to hasten the looked-for day."[68]

Following the president, administration officials praised the agency established by the statute as reflecting the vision of the president and as a useful addition to the UN family. They argued that throughout the negotiations, the United States had not sacrificed any element of substance or principle. According to Bechhoefer, "at all stages . . . the negotiations adhered strictly to the terms of reference contained in the [Atoms for Peace] address."[69] And, in the view of Ambassador

James Wadsworth, the negotiations on the statute produced a document that:

> will make possible an Agency with broad authority to assist in research and development in the peaceful uses field; possess and distribute nuclear materials; carry out the pooling of such materials at the request of member states as proposed by the President; establish and operate its own facilities; organize and apply a system of minimum safeguards on request to bilateral or multinational arrangements or the atomic energy activities of a member state; conduct its financial management on a flexible but business-like basis in the interest of the entire membership; establish an appropriate relationship with the United Nations and other international organizations; and take into consideration recognized standards of international conduct in connection with the admission of new members.[70]

Nonetheless, the international pool, so central to the original concept, never fully materialized and, as a consequence, never played the role envisioned for it in the president's proposal. In Hewlett's view, the "President's several commitments of fissionable material to the international agency did not stimulate significant responses from other nations, and the announcements were taken for what they were, more a gesture than a significant step toward reducing the dangers of proliferation."[71] Indeed, from ownership and control over nuclear materials, a "clearinghouse" function for the agency slowly and hesitantly but inexorably emerged. As for safeguards, the concept accepted was regarded as minimal by the United States; it was considerably less than that thought necessary by the majority during discussions in the UNAEC in the late 1940s, but then the whole basis for nuclear development had changed to one predicated on national programs and structures.

If the child did not fully live up to expectations, part of the blame can be attached to its parent and part to changes that had come about in the environment between the Atoms for Peace speech and the *demarrage* of the agency. In the view of some observers, the IAEA was undermined by U.S. actions, in particular by the U.S. campaign to promote EURATOM and bilateral agreements for cooperation; by the U.S. failure to require the Europeans to accept agency safeguards, because EURATOM safeguards were deemed sufficient; and by U.S. insistence that the agency's first director general be American, a demand that reinforced early Soviet concerns that the agency would be a servant of U.S. interests and policy.[72] Not all would agree with this sharp assessment. On the matter of bilateral agreements, for example, it has been argued that the "greatest threat to the agency was that the entire program for peaceful use of the atom might lag" and that the bilaterals actually assisted the establishment of the agency.[73]

The environmental changes were equally severe. When President Eisenhower laid his proposal before the world, both source and special nuclear material were in short supply and expected to remain so for a long time. Information on nuclear energy was also very limited, owing largely to the U.S. policy of secrecy in the first postwar decade; and an expectation was setting in that competitive and economic nuclear power was imminent. By the time the agency was under way in 1957, much had changed. There was a glut of uranium in the market; information was widely and abundantly available because of the 1955 United Nations Conference on the Peaceful Uses of Atomic Energy; and the bloom was off the rose insofar as early deployment of cheap and economic nuclear energy was concerned.

In short, economic and technological conditions on the one hand, and the policy pursued by the leading nuclear power on the other, limited the scope as well as the importance of agency activity even before the IAEA was under way. But an intellectually resilient and entrepreneurial secretariat, with support from the United States, would piece together a program of activities that would set the agency on the way to tackling the problem of the peaceful development of atomic energy.

NOTES

1. "Joint Declaration by the Heads of Government of the United States, the United Kingdom, and Canada, November 15, 1945," in U.S. Department of State, Historical Office, Bureau of Public Affairs, *Documents on Disarmament 1945–1959*, Pub. No. 7008, 2 vols. (Washington, D.C., U.S. Government Printing Office, 1960) vol. I., p. 1.

2. Ibid.

3. Ibid., p. 2.

4. Ibid.

5. "Moscow Communiqué by the Foreign Ministers of the United States, the United Kingdom, and the Soviet Union (Extracts), December 27, 1945," ibid., p. 5.

6. For a discussion of the establishment of the committee and its board of consultants, and a recounting of the preparation of the report, see, Richard

G. Hewlett and Oscar E. Anderson, Jr., *A History of the United States Atomic Energy Commission*, vol. I, *The New World, 1939–1946* (Washington, D.C., U.S. Atomic Energy Commission, 1972) pp. 531–554.

7. "The Baruch Plan: Statement by the United States Representative [Baruch] to the United Nations Atomic Energy Commission, June 14, 1946," in *Documents on Disarmament* (see note 1 above) p. 10.

8. Ibid., pp. 10–11.

9. Ibid., p. 11.

10. Ibid.

11. "Address by the Soviet Representative [Gromyko] to the United Nations Atomic Energy Commission, June 19, 1946," ibid., p. 18.

12. Ibid.

13. "First Report of the United Nations Atomic Energy Commission to the Security Council (Extract), December 31, 1946," ibid., pp. 50–59.

14. See, for example, "Soviet Proposals Introduced in the United Nations Atomic Energy Commission, June 11, 1947," ibid., pp. 85–88.

15. See, "Minutes of Meeting at the Department of State, April 2, 1948, on International Control of Atomic Energy, Department of State Atomic Energy Files," (pp. 318–322) and Part I, and "Position Paper Approved by the Executive Committee in Regulation of Armaments, April 9, 1948, United States Policy in the United Nations for the International Control of Atomic Energy, Department of State Atomic Energy Files, RAC D-301a" (pp. 323–335) reprinted in vol. 1, *Foreign Relations of the United States, 1948, Vol. I, General; The United Nations*, in 2 parts (Washington, D.C., U.S. Government Printing Office, 1975).

16. For a fuller discussion of these areas of disagreement, see "United States Policy in the United Nations for the International Control of Atomic Energy, " ibid., pp. 328–329.

17. "Third Report of the United Nations Atomic Energy Commission to the Security Council, May 17, 1948," in *Documents on Disarmament* (see note 1 above) pp. 171–172.

18. Ibid., p. 171.

19. Ibid,. pp. 170–171.

20. P.L. 79–585, 60 Stat. 755.

21. See Richard G. Hewlett and Francis Duncan, *A History of the United States Atomic Energy Commission*, vol. II, *Atomic Shield, 1947–1952* (Washington, D.C., U.S. Atomic Energy Commission, 1972) pp. 479–484.

22. P.L. 82–235, 65 Stat. 692.

23. "United States 'Atoms-for-Peace' Proposal: Address by President Eisenhower to the General Assembly, December 8, 1953," in *Documents on Disarmament* (see note 1 above) pp. 399–400.

24. Ibid., p. 395.

25. Ibid., pp. 395–396.

26. Ibid., p. 396.

27. Ibid., p. 398.

28. Ibid.

29. Ibid.

30. Safeguards concepts are discussed in detail in chapters 4 and 5.

31. James R. Schlesinger, "Atoms for Peace Revisited," in Joseph F. Pilat, Robert E. Pendley, and Charles K. Ebinger, eds., *Atoms for Peace: An Analysis After Thirty Years* (Boulder, Colo., and London: Westview Press, p. 5.

32. Richard G. Hewlett, "From Proposal to Program," ibid., p. 28.

33. "Statement by the Soviet Government on President Eisenhower's Atoms-for-Peace Address, December 21, 1953" in *Documents on Disarmament* (see note 1 above) pp. 404–405.

34. Ibid., p. 405.

35. Ibid., p. 406.

36. On the origins of Atoms for Peace, see, for example, Henry David Sokolski, *Eisenhower's Original Atoms-for-Peace Program: The Arms Control Connection*, Occasional Paper No. 52 (Washington, D.C., The Wilson Center, International Security Studies Program, 1983).

37. Remarks made by Cole during discussions at a conference on "Atoms-for-Peace: After Thirty Years," sponsored by the Center for Strategic and International Studies, Georgetown University, and the Los Alamos National Laboratory, Washington, D.C., December 7–8, 1983.

38. Jack M. Holl, "The Peaceful Atom: Lore and Myth," in Pilat, Pendley, and Ebinger, eds., *Atoms for Peace* (see note 31 above) pp. 152–154.

39. Ibid., p. 152.

40. Hewlett (see note 21 above) pp. 32–33.

41. "Message from President Dwight D. Eisenhower to the Congress 'Recommendations Relative to the Atomic Energy Act of 1946,' February 17, 1954," reprinted in *Atoms for Peace Manual: A Compilation of Official Materials on International Cooperation for Peaceful Uses of Atomic Energy, December 1953–July 1955*, Senate Document No. 55, 84 Cong. 1 sess. (1955) (Washington, D.C., U.S. Government Printing Office, 1955) p. 9.

42. According to one analyst, Eisenhower regarded the legislative changes he recommended as separate from his Atoms-for-Peace proposal, additional legislation for which would be considered as required on the basis of negotiations with other nations. However, subsequent legislation was not required. See Warren H. Donnelly, *Commercial Nuclear Power in Europe: Interaction of American Diplomacy with a New Technology*, prepared for the Subcommittee on

National Security Policy and Scientific Developments of the House Committee on Foreign Affairs, December 1972 (Washington, D.C., U.S. Government Printing Office, 1972) p. 23.

43. Bernard G. Bechhoefer, "Negotiating the Statute of the International Atomic Energy Agency," *International Organization* (Winter 1959) p. 41.

44. Hewlett (see note 32 above) p. 28.

45. Ibid.

46. "Memorandum: Outline of an International Atomic Energy Agency," in *Atoms for Peace Manual* (see note 41 above) p. 267.

47. Ibid., pp. 268–269.

48. Bechhoefer (see note 43 above) p. 42.

49. Ibid., pp. 43–44.

50. "Memorandum from the Assistant Secretary of State for European Affairs [Merchant] to the Soviet Ambassador [Zaroubin], July 9, 1954," in *Documents on Disarmament* (see note 1 above) p. 428.

51. Ibid.

52. Ibid., p. 429.

53. Ibid.

54. "Soviet Proposals Introduced in the United Nations Atomic Energy Commission, June 11, 1947" (see note 14 above) p. 86. On the significance of this proposal, see Bertrand Goldschmidt, "A Forerunner of the NPT? The Soviet proposals of 1947," *IAEA Bulletin* vol. 28, No. 1 (Spring 1986) pp. 58–64.

55. Report of this meeting is based on information provided by the U.S. participant, Professor I. I. Rabi to Dr. John Hall who served as a senior advisor to the author in the early stages of the preparation of this book.

56. See Bechhoefer (see note 43 above) pp. 45–46.

57. This account of the negotiations on the international statute of the agency is derived from the "Progress Report on International Atomic Energy Agency Negotiations," which was submitted by Morehead Patterson, U.S. Representative for International Atomic Energy Agency Negotiations to the President on November 30, 1955, in *Department of State Bulletin*, January 2, 1956, p. 6.

58. For a full and lively discussion of the Geneva Conference, see Bertrand Goldschmidt, *The Atomic Complex: A Worldwide Political History of Nuclear Energy* (La Grange Park, Ill., American Nuclear Society, 1982) pp. 257–262.

59. Bechhoefer (see note 43 above) p. 50.

60. The following discussion relies on a report by James T. Ramey (executive director) and George Norris, Jr. (committee counsel) on the Conference of the Statute of the International Atomic Energy Agency (November

27, 1956) at the Sterling-Cole Collection, Colgate University Library, Hamilton, N.Y., and on review of the records of the proceedings.

61. For a fuller discussion of the safegaurds debate by a participant at the conference, see Goldschmidt (note 58 above) pp. 281–283.

62. See the *Statute of the International Atomic Energy Agency*, hearings before the Senate Committee on Foreign Relations and Senate Members of the Joint Committee of Atomic Energy, on Executive 1, 85 Cong. 1 sess. 1957 (Washington, D.C., U.S. Government Printing Office, 1957) pp. 4–5.

63. Ibid., pp. 14–15.

64. Ibid., pp. 49, 92. See also Donnelly (see note 42 above) p. 51.

65. P.L. 85–177, 71 Stat. 453.

66. "Statement of President Dwight D. Eisenhower at the Closing Session of the Conference on the Statute of the International Atomic Energy Agency," White House Press Release dated October 21, 1956, in *Department of State Bulletin*, November 19, 1956, p. 814.

67. Ibid., p. 815.

68. Ibid.

69. Bechhoefer (see note 43 above) p. 40.

70. *Statute of the International Atomic Energy Agency* (see note 62 above) p. 46.

71. Hewlett (see note 32 above) p. 30.

72. See, for example, ibid.

73. Bechhoefer (see note 43 above) p. 52.

Chapter 3

THE IAEA: ITS STRUCTURE AND ACTIVITIES

Unlike many other international organizations created by or affiliated with the United Nations, the International Atomic Energy Agency (IAEA) is an autonomous agency with its own statute and board of governors. Although its statute provides that it "conduct its activities in accordance with the purposes and principles of the UN to promote peace and international cooperation" and the director general submits an annual report of its activities to the UN General Assembly, it is *not* a specialized agency of the UN and it acts independently of the UN.[1]

The principal components of the IAEA consist of the membership as a whole and its general conference, the board of governors, and the agency secretariat headed by the director general. Of the three components, the board of governors, as the executive organ, is the most important for the operation of the agency.

THE BOARD OF GOVERNORS

President Eisenhower submitted the statute of the agency to the Senate on March 21, 1957, for its advice and consent to ratification, saying that "Authority for directing the Agency will rest primarily in a Board of Governors. The method of choosing those Governors was considered with particular care. The formula finally agreed upon balances geographic considerations with the capacity of the cooperating nations to supply technical or material support to Agency projects."[2] The board of governors, by statute, was to comprise those countries most advanced in the technology of atomic energy as well as countries that were producers of source material. In the international debate on the statute when it was first proposed, there was considerable criticism by countries whose nuclear capacities were small of the dominant position of the atomically advanced countries on the board. However, the twelve sponsoring countries (Australia, Belgium, Brazil, Canada, Czechoslovakia, France, India, Portugal, South Africa, USSR, UK, and United States) of the draft that served as the basis for the agency statute (all of which were either producers of source material or were nuclear advanced) held firm. The board was to consist of twenty-three members—five representing the countries "most advanced in the technology of atomic energy including the production of source material" and five more members from countries "most advanced in the technology of atomic energy" in the following areas not represented by the aforementioned five: North America, Latin America, Western Europe, Eastern Europe, Africa and the Middle East, South Asia, Southeast Asia and the Pacific, and the Far East. In addition, two members were to be selected from among the following producers of source materials: Belgium, Czechoslovakia, Poland, and Portugal, and one other member as a supplier of technical assistance. These thirteen board members were designated by the outgoing board (in the case of the first board, by the preparatory commission, which was made up of the twelve sponsoring countries and Argentina, Japan, Egypt, Peru, Indonesia, and Pakistan). An additional ten board members were to be elected by the general conference with due regard for equitable geographical representation.[3]

The size and composition of the board of governors has been changed three times by amendments to the statute. On January 31, 1963, an amendment increased the representation from Africa and the Middle East, resulting in a total board membership of twenty-five. A more significant increase in the board's membership was made, effective June 1, 1973, when the statute was amended to enlarge the board to thirty-four. The globally most advanced member seats were increased from the original five to nine, and the number of elected members was increased from eleven to twenty-two. The most recent change in board membership was voted in 1984 to make room for the People's Republic of China by further expanding the number of globally most advanced seats on the board.[4] Thus, at present, each outgoing board of governors designates for membership on the next board the ten members most advanced in the technology of atomic energy and the member most advanced in any of eight specified geographic regions for which there is no globally most advanced member on the board. In addition, twenty-two other members are elected on a basis of geographic distribution to the board by the general conference, eleven each year for two-year terms.

The board of governors usually meets four times a year, generally in February, June, and immediately before and after the September general conference. As the executive organ of the agency, it acts on agency membership applications, budgets, programs, and projects, and approves all safeguards arrangements and safety standards. The board of governors also has responsibility for appointing, with the approval of the general conference, the director general of the agency.

THE GENERAL CONFERENCE

The general conference of the IAEA meets only once a year—generally in Vienna in September—for general debate on agency programs and to hear and consider for approval the director general's annual report, the agency budget for the forthcoming year, applications for new membership in the agency, and to elect those members to the board of governors that are its statutory responsibility. On several occasions, in response to invitations from member governments, the general conference has been held outside of Vienna: in 1965, Tokyo hosted the general conference; in 1972, it met in Mexico City; in 1976, in Rio de Janeiro; and in 1979, in New Delhi. Fifty-nine member states participated in the first general conference of the IAEA; the current agency membership is 113.

The general conference has authority to make decisions on matters referred to it by the board or "to propose matters for consideration by the Board and to request from the Board reports on any matters relating to the functions of the Agency."[5] This authority has been frequently invoked, and the limited influence that the conference does have on agency policy and orientations stems from resolutions passed under this provision.[6] Although the conference has the power to return the budget to the board with recommendations for change, it has never used this authority. The true value of the conference lies not in its limited formal powers, but in its role as a forum where those not represented on the board can express their views and exchange information, and where member states can finalize bilateral or multilateral agreements in the nuclear field. As will be discussed in chapter 7, its role as a forum has had the unfortunate effect of providing a vehicle for introducing extraneous political issues into agency deliberations.

THE SECRETARIAT

The Secretariat, headed by the director general, is the operating organ of the IAEA. The director general, appointed by the board of governors and confirmed by the general conference for a four-year term, is the chief administrative officer of the agency. In its 28-year history, the agency has had only three directors general. The first, U.S. Congressman Sterling Cole, served four years. He was succeeded by the Swedish nuclear scientist Sigvard Eklund, who served between 1961 and 1982. The current director general, Hans Blix, is a Swedish diplomat. The director general is assisted by five deputies, each heading a department. Deputy directors general head the departments of administration, technical cooperation, nuclear energy and safety, research and isotopes, and safeguards.

Patterns developed early with respect to staffing the key positions. Thus, ever since the departure of Sterling Cole as director general in 1961, the deputy director general for administration (who has staffing, external relations, and a general oversight responsibility) has been an American. The technical cooperation post, established in 1964, has been filled by a national from the Third World; research and isotopes by a national from an Organization for Economic Cooperation and Development (OECD) nation; nuclear energy and safety, by a Soviet national; and the safeguards post has been occupied by an Australian, two Swiss, and a German.

The director general has two important advisory bodies—the Scientific Advisory Committee (SAC) and the Standing Advisory Group on Safeguards Implementation. In addition, he convenes panels of experts for advice and recommendations for specific purposes.

The agency staff, as of the end of 1985, numbered 1,942, of whom 715 were in the professional or higher category, 1,091 were in the general services category, and 136 in the maintenance and service category. All posts in the professional and higher category, with the exception of those requiring special linguistic qualifications, are filled on the principle of geographic distribution. Posts in the general services category are generally subject to local recruitment.

The initial agency staff, at the end of its first year of operation, comprised a total of only 130 professionals, directors, and the deputy directors general.[7] Thirty-five of these posts were filled by interpreters and translators. By the next year, the approved manning table of the secretariat comprised a total of 220 posts at the level of professional and higher, of which forty-one were linguists.[8]

THE IAEA BUDGET

The budget of the agency has grown appreciably over the life of the IAEA, but is still modest by comparison with other international organizations and member agencies in the United Nations family. The IAEA budget falls into two categories. One component is the regular budget and the other is the operational budget. The regular budget provides for the ordinary administrative expenses of the agency as well as for expert panels, symposia and conferences, special missions, and information services. Safeguards expenses are included in the regular budget, although a special formula, designed to limit costs to developing countries, is used for purposes of assessment.[9] The regular budget is funded by contributions made according to a formula of annual assessments on each member state and by miscellaneous income.[10] The operational budget, financed by voluntary contributions paid into a general fund established for this purpose, is used to fund the agency's laboratory and research projects (discussed below) and for technical assistance, training, and research contracts. In addition, the agency receives voluntary in-kind contributions from member states.

The IAEA regular budget for 1960 was $5.843 million; in 1965 it was $7.938 million; in 1975 the regular budget was $29.675 million. The regular budget total for 1986 was $98.680 million, of which $90.570 million was from assessed contributions of member states, $3.704 million from income from work for others, and $4.406 million from other miscellaneous income. The regular budget for 1987 is $103.899 million. This represents virtually no real growth over the preceding year. Indeed, since 1983, the agency has operated under the principle of zero real growth and has had to pay the increased costs of some programs, such as safeguards, by curtailing activities in other programs or by relying on extrabudgetary contributions. Most of the additional costs of the supplementary safety program[11] developed after the Chernobyl accident were funded by the reordering of some programs in the regular budget as well as by voluntary contributions. Table 3-1 gives a breakdown of the 1986 regular budget by program area and programs.

Voluntary Contributions

In 1959, the target for voluntary contributions to the operational budget was $1.5 million; in each of the years from 1962 through 1970, the target for voluntary contributions was set at $2.0 million. These targets, however, were never reached; it was not until 1974 when the target had grown to $3.0 million, that it was reached. Since 1974, the target has increased yearly, and by 1984 was $22.5 million. Table 3-2 and figure 3-1 illustrate the growth of the technical assistance and cooperation fund during the past decade. In 1985, the

TABLE 3-1. THE AGENCY'S BUDGET FOR 1986
(dollars)

Program area/program	Regular budget estimates, 1986
1. Nuclear power and the fuel cycle:	
1.1. Nuclear power planning and implementation in developing countries	1,434,000
1.2. Nuclear power plant performance	1,066,000
1.3. Nuclear fuel cycle	1,455,000
1.4. Radioactive waste management	2,838,000
1.5. Advanced systems and applications	1,341,000
Subtotal	8,134,000
2. Nuclear applications:	
2.1. Food and agriculture	2,994,000
2.2. Human health	2,330,000
2.3. Physical sciences and technology	3,764,000
2.4. The laboratory	4,247,000
2.5. International Center for Theoretical Physics	1,170,000
Subtotal	14,505,000
3. Nuclear safety and radiation protection:	
3.1. Radiation protection	2,141,000
3.2. Safety of nuclear installations	2,337,000
3.3. Risk assessment	523,000
Subtotal	5,001,000
4. Safeguards:	
4.1. Safeguards implementation	20,457,000
4.2. Safeguards development and support	12,884,000
Subtotal	33,341,000
5. Direction and support area:	
5.1. General management and secretariat of the policy-making organs	5,877,000
5.2. Administration	7,150,000
5.3. Technical cooperation servicing and coordination	5,022,000
5.4. General services	9,981,000
5.5. Specialized service activities	5,074,000
5.6. Shared support services	891,000
Subtotal	33,995,000
Total agency programs	94,976,000
Services provided to others	3,704,000
Total	98,680,000
Source of funds:	
Assessment on member states	90,570,000
Income from work for others	3,704,000
Other miscellaneous income	4,406,000

Source: IAEA, *The Agency's Budget for 1986* GC(XXIX)/750, August 1985.

TABLE 3-2. TECHNICAL ASSISTANCE AND COOPERATION FUND, 1976–1985
(dollars)

Program year	Target for voluntary contributions to the Technical Assistance and Cooperation Fund	Amount pledged[a]	Amount actually made available for technical cooperation by program year[a]
1976	5,500,000	5,061,957	5,492,167
1977	6,000,000	5,449,466	5,962,688
1978	7,000,000	6,451,332	7,121,508
1979	8,500,000	8,062,513	8,802,221
1980	10,500,000	10,059,733	10,632,033
1981	13,000,000	12,053,611	12,955,595
1982	16,000,000	14,901,346	16,003,198
1983	19,000,000	17,619,372	19,244,903
1984	22,500,000	20,735,931	22,231,347
1985	26,000,000	23,255,051	25,193,932

Source: IAEA, *The Agency's Technical Cooperation Activities in 1985*. Report of the Director General GC(XXX)/INF/234, August 1986, p. 62.

[a]These amounts include "additional income" over and above the amounts pledged.

voluntary contributions of member states and such miscellaneous income as interest earnings and payments of program costs by recipient countries (making up the Technical Assistance and Cooperation Fund—TACF) was $25.1 million. The total resources available in 1985 in the operational budget was $38.1 million. This was made up of the $25.1 million in TACF, $2.6 million through the United Nations Development Programme (UNDP), $7.8 million in extrabudgetary funds, and $2.7 million as assistance in-kind.[12] Figure 3-2 shows the growth in resources available for the agency technical cooperation programs from 1979 through 1985.

THE SCIENTIFIC ADVISORY COMMITTEE

In 1954, the United Nations General Assembly, recognizing the secretary-general's need for scientific advice on arrangements for the First International Conference on the Peaceful Uses of Atomic Energy, scheduled to take place the following year in Geneva, passed a resolution creating a United Nations Scientific Advisory Committee (SAC). The UN SAC, made up of a small group of distinguished scientists (usually from the government's nuclear sector), was charged with advising the secretary general on matters relating to the peaceful uses of atomic energy with which the United Nations might be concerned. The committee was chaired by the UN secretary general.

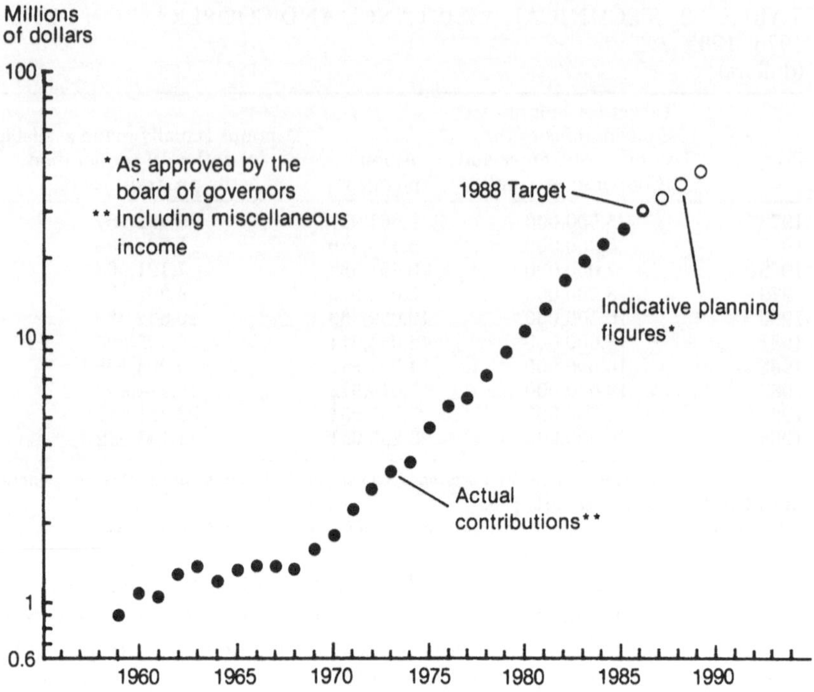

Figure 3-1. Technical Assistance and Cooperation Fund voluntary contributions. *Source*: IAEA, *The Agency's Technical Co-operation Activities in 1985*, Report by the Director-General, GC(XXX)/INF/234, August 1986, p. 62.

When the IAEA was established, its board of governors recognized the desirability of setting up a similar committee to give scientific advice to the director general. Since the mandates of both scientific advisory committees were similar, the membership of the IAEA SAC, appointed in November 1958 by the board of governors on the basis of recommendations submitted by the director general, was identical to that of the UN SAC and remained so for a number of years. The scientists appointed to the first IAEA SAC were:[13]

- Homi Bhabha, an Indian physicist who presided over the 1955 Geneva Conference and was chairman of the Indian Atomic Energy Commission.
- John Cockcroft, an English Nobel laureate in physics.
- Vasily Emelyanov, a Russian engineer and member of the USSR Academy of Sciences.
- Bertrand Goldschmidt, a French chemist and director of the

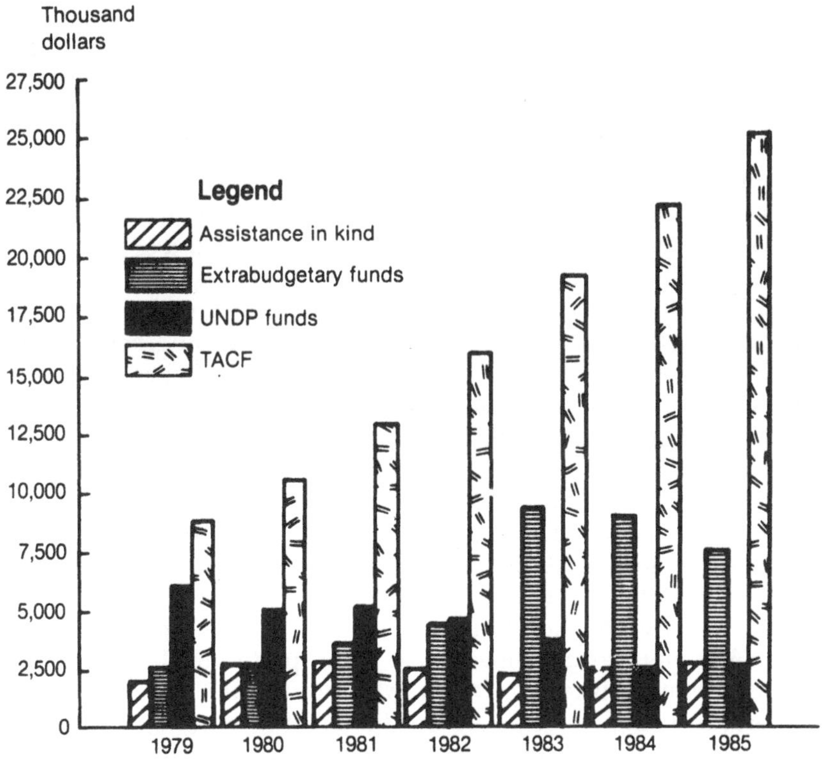

	1979	1980	1981	1982	1983	1984	1985
TACF	8,802	10,632	12,956	16,003	19,241	22,232	25,197
Extra budgetary funds	2,635	2,669	3,525	4,413	9,394	9,062	7,484
Assistance in kind	2,015	2,628	2,788	2,493	2,172	2,066	2,765
UNDP	6,066	5,018	5,186	4,631	3,706	2,541	2,654
Total	19,518	20,947	24,455	27,540	34,513	35,901	38,100

Figure 3-2. Resources available for agency technical cooperation programs, 1979–1985. *Source*: IAEA, *The Agency's Technical Co-operation Activities in 1985*, Report by the Director-General, GC(XXX)/INF/234, August 1986.

French Commisariat de l'Energie Atomique Chemistry Department and the Department of External Affairs.

- Bernhard Gross, the head of the Brazilian National Institute of Technology's Department of Electrical Measurements.
- W. Bennett Lewis, a Canadian physicist and vice president for research of Atomic Energy of Canada, Ltd.

- Isidor I. Rabi, an American Nobel laureate in physics and chairman of the President's Science Advisory Committee and of the U.S. Atomic Energy Commission's General Advisory Committee.

The chairman of the IAEA SAC was the director general, but the committee's technical discussions were presided over by its vice chairman. The function of the IAEA SAC was, and continues to be, to provide the director general and through him the board of governors with scientific advice on questions relating to the agency's activities. At the first two sessions of the IAEA SAC, members discussed the agency's scientific program; considered proposed conferences, symposia, and seminars to be sponsored by the agency; shaped the agency's scientific publications program; determined research contracts awarded or to be awarded; and reviewed the agency's contracting policies in general. At these first sessions, the committee also addressed the scientific considerations involved in the draft regulations for the application of international safeguards by the IAEA. The committee also considered a project to exchange information on fusion (following up on the 1958 Geneva conference).

The members of the IAEA SAC, having worked together at the UN SAC, and having known each other professionally for many years, had no difficulty working with one another at the IAEA. (Several of them, in fact, were on the board of governors together as well.) The attitude of the members was best described and explained by Isidor Rabi, who, in an address at the Salazar Atomic Center in Mexico in October 1972, said:

> The universal nature of science and the character of the scientific community, which shares this universality, made it possible to have close and fruitful relations between scientists of different cultures and even of countries of seemingly antagonistic political orientations. Our committee became like a club where the members learned to respect and trust one another. Our trust was never misplaced or betrayed. As a result, the committee could do its work with the utmost expedition and objectivity to the astonishment of experienced and sophisticated diplomats.[14]

During Sigvard Eklund's long tenure as IAEA director general, the Scientific Advisory Committee played an active and useful role in agency affairs. Himself a scientist, Eklund had excellent relations with the committee members and he used the committee effectively. He expanded its membership and, despite opposition from a number of Moslem and other countries, he appointed a highly respected

Israeli scientist, the physical chemist Israel Dostrovsky, to the IAEA SAC. In the early 1970s, the scientists on the IAEA committee were Goldschmidt (France), Lewis (Canada), Gerald Tape (United States), Walter Marshall (UK), Homi Sethna (India), Ivan Chuvilo (USSR), Hervasio de Carvalho (Brazil), Wolf Haefele (Federal Republic of Germany), Tamaki Ipponmatsu (Japan), Bruno Straub (Hungary), M.A.N. El-Guebeily (Egypt), and Dostrovsky (Israel). By that time, only a few of the IAEA committee members were also members of the UN SAC. While the IAEA committee remained active, the UN committee's role diminished in time. Rabi noted, in his 1972 Salazar Center speech that "unfortunately . . . the utilization of the mechanism of the UN SAC has not been followed by succeeding Secretary Generals of that body."[15] The IAEA SAC, however, continued to play a significant role. Rabi found that "under the beneficent leadership of Dr. Eklund and supported by his Scientific Advisory Committee and the Board of Governors, the Agency has assumed the role of Science Advisor to member states," and he visualized for the future "an expanding role for the Agency as a global Science Advisor to all nations."[16]

When Hans Blix replaced Eklund as IAEA director general, he adjusted the character of the IAEA SAC, replacing some scientists on the committee (such as Dostrovsky, Goldschmidt, Haefele, Marshall, and Sethna) with administrators and engineers to reflect changes in the nuclear environment and in emphasis in agency tasks and responsibilities. An agency press release in September 1982 announcing the new membership of the SAC stated that "compared with the outgoing Committee, the new Committee includes a larger number of experts in nuclear safety and waste management who are responsible in their national governments for regulating nuclear industry. This shift in balance reflects the growing importance of the IAEA's program for ensuring the safety of nuclear power and for dealing with problems of waste management."[17]

THE AGENCY'S PROGRAMS

The objectives of the IAEA, as succinctly stated in Article II of the agency statute, are that "the Agency shall seek to accelerate and enlarge the contribution of atomic energy to peace, health and prosperity throughout the world. It shall insure, so far as it is able, that assistance provided by it or at its request or under its supervision or control is not used in such a way as to further any military purpose."

The assistance provided its member countries covers the entire spectrum of applied and basic nuclear science, from radioactive isotopes for industrial applications to biology and medicine, from nuclear power to theoretical physics. An idea of the scope of the agency's activities and responsibilities may be gathered from its organizational structure. This is shown in the agency's organizational chart (figure 3-3). Experts provide direct technical assistance, which includes material and equipment, fellowships, training courses, and workshops. Technical assistance is also made available through the agency laboratories, its publications, symposia, etc.

In 1985, assistance in the field of agriculture accounted for almost one-quarter of the technical assistance budget. This was followed, in order of support, by nuclear technology, nuclear safety, isotope and radiation applications in industry and hydrology, and nuclear physics. Table 3-3 shows the allocation of agency resources for 1983 and 1984 by field of activity and type of assistance. Figure 3-4 gives similar data on resource utilization for 1985 in a newly adopted agency format. Figure 3-5 shows the distribution of technical cooperation in 1985 by field of activity and geographic region. Over the years, the degree of agency support for the various fields of technical activity has changed both as the needs of the member countries have changed and as the technical requirements and, in some cases (such as peaceful nuclear explosives), the economic and political justifications have changed. Although most of the agency's programs of assistance are aimed at the developing countries (which make up the overwhelming majority of the IAEA membership), some programs, such as the agency's International Nuclear Information System (INIS), are of particular benefit to the advanced industrial member countries.[18]

The IAEA cooperates with a number of other international and regional agencies in developing and carrying out programs and providing assistance in specific areas of research and technology. Thus, in applying nuclear techniques to food and agriculture, the agency works very closely with the UN Food and Agriculture Organization (FAO) and, as shown in the IAEA organization chart (figure 3-3), operates a joint division of Isotope and Radiation Applications of Atomic Energy for Food and Agricultural Development. Over 400 projects in member states are supported by the joint IAEA/FAO division each year. These include programs concerned with soil fertility, irrigation, and crop production using radioisotope tracers. Work in the field of mutation plant breeding and genetics is also being conducted under a number of research programs.

The program on insect control by the sterile insect technique, started in the early days of the agency, continues to show remarkably successful results. Using this technique, insects are sterilized by radiation in the laboratory and then released in great numbers to mate with native flies, thus producing no offspring, and resulting in a rapid fall in the native insect population over a comparatively small number of generations. First used to effectively control the screwworm, the technique has subsequently been used in controlling the Mediterranean fruit fly, and research is underway to apply this technique to the tsetse fly. Another program of the Food and Agricultural Development Division, which has been supported since the early days of the agency, is the preservation of food by radiation. By intense irradiation, food can be sterilized and preserved for long periods of time. A more common application of food irradiation is the reduction of insect infestation and the reduction of spoilage-causing bacteria, thereby increasing the shelf-life of the irradiated product and reducing food spoilage in transport. This program is continuing with the added cooperation of the World Health Organization (WHO). In 1978, under an agreement between the IAEA, FAO, WHO, and the government of the Netherlands, an International Facility for Food Irradiation Technology was established in Wageningen, the Netherlands.

The IAEA has an active program of research and assistance in biology and the medical sciences. Many of its programs are carried out in cooperation with WHO. Radiation techniques and radioactive isotopes are used in basic biomedical research, for therapy in such diseases as cancer and, in trace quantities as a diagnostic tool in medicine. The agency supports programs in all these areas. Radiation is also being applied under agency programs in parasite control. The sterilization of medical equipment and medical supplies by irradiation has also been demonstrated to be an effective technique. Many of the aforementioned applications of radiation and radioisotopes require accurate radiological measurements, dosimeters, and dosimetry standards. A joint IAEA-WHO network of secondary standard dosimetry laboratories has been set up and currently has forty-eight participating laboratories. The agency's own dosimetry laboratory serves as the central facility.

The IAEA program in the physical sciences covers the range from highly theoretical fundamental research in physics and chemistry to applied technology, nuclear engineering, and hydrology. Theoretical studies are carried out at the International Center for Theoretical Physics in Trieste, Italy. The programs of this center, as well as of the several IAEA research laboratories, will be discussed below. The agency's programs in applied technology include work on industrial applications and training programs on the use of nuclear techniques

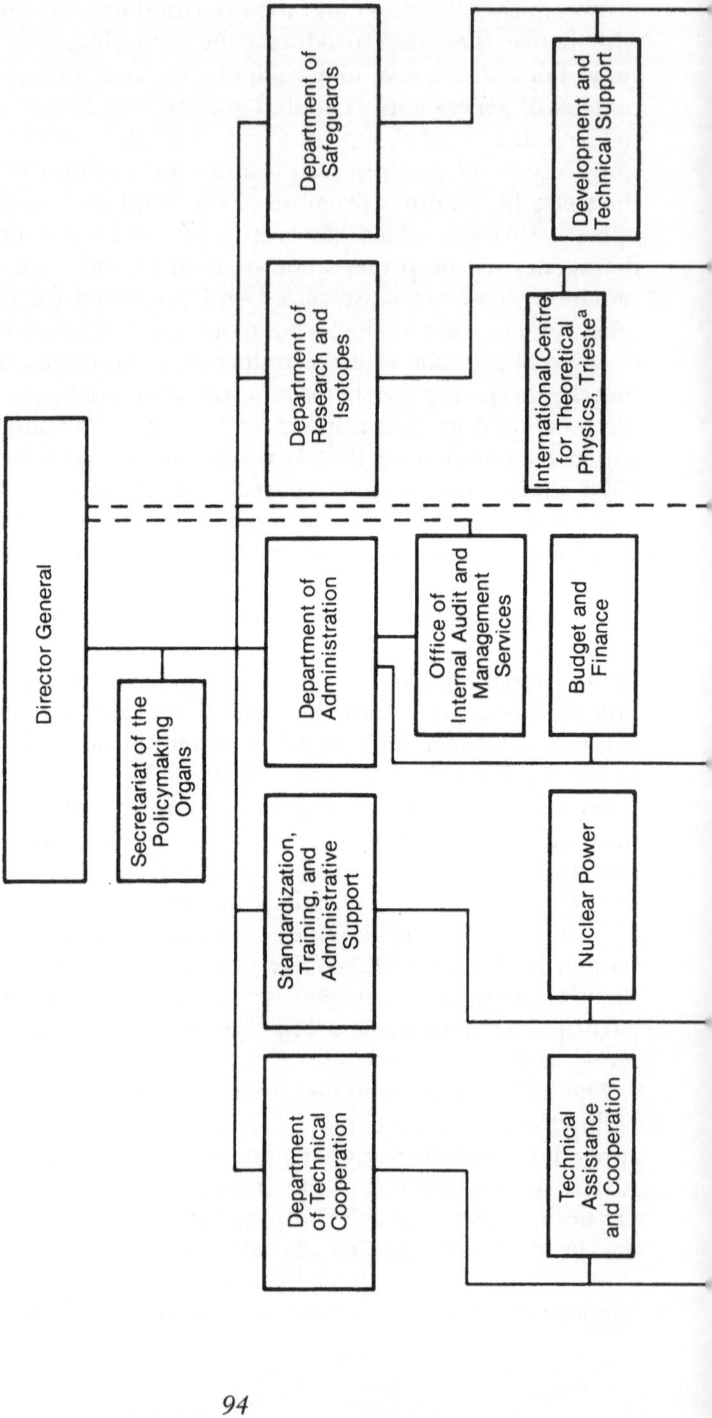
Director General
Secretariat of the Policymaking Organs
Department of Safeguards
Development and Technical Support
Department of Research and Isotopes
International Centre for Theoretical Physics, Trieste[a]
Department of Administration
Office of Internal Audit and Management Services
Budget and Finance
Standardization, Training, and Administrative Support
Nuclear Power
Department of Technical Cooperation
Technical Assistance and Cooperation

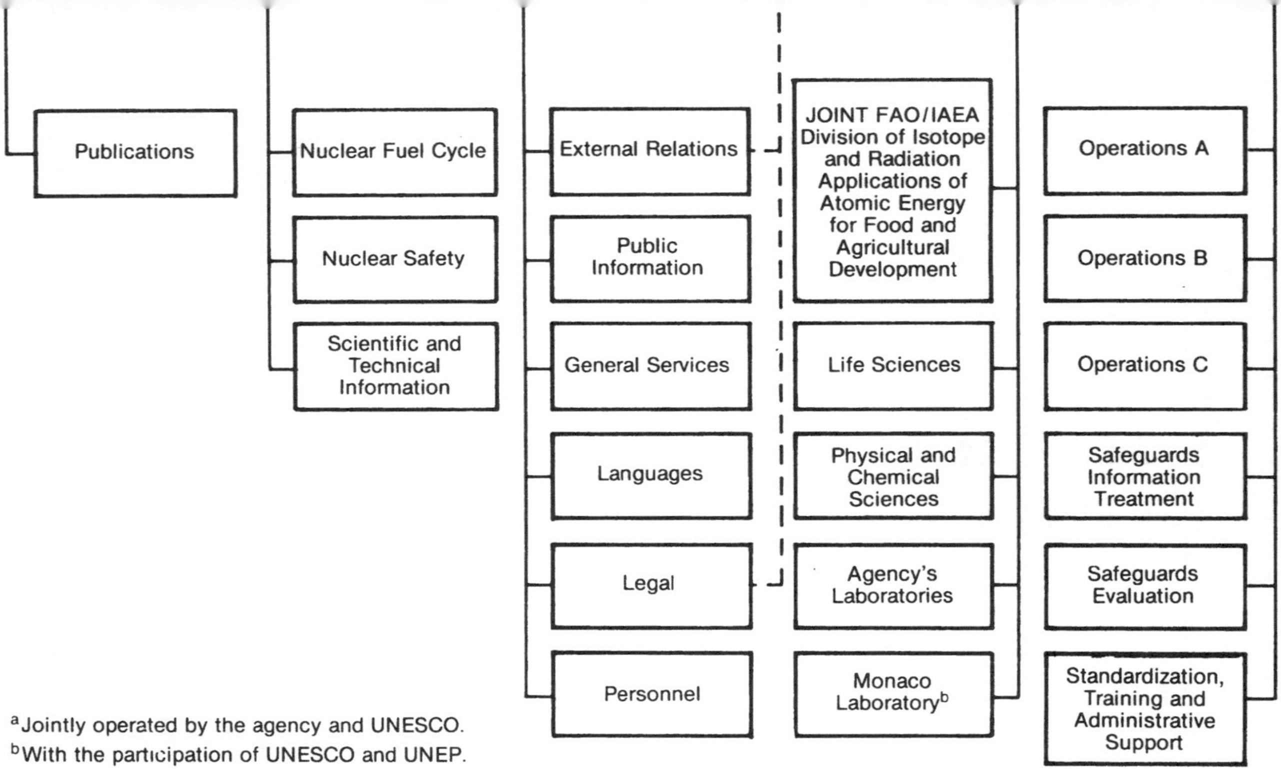

[a]Jointly operated by the agency and UNESCO.
[b]With the participation of UNESCO and UNEP.

Figure 3-3. Organizational chart. *Source*: IAEA, *The Annual Report for 1985*, GC(XXX)/775, July 1986.

TABLE 3-3. UTILIZATION OF RESOURCES, 1983 AND 1984
(absolute numbers in thousands of dollars)

Field of activity	Year	Experts	Equipment	Fellow-ships		Share of total program (%)
General atomic energy development	1983	642.9	1,123.9	383.9	2,150.7	8.1
	1984	857.6	1,570.0	342.9	2,770.5	8.5
Nuclear physics	1983	381.4	1,878.0	346.2	2,605.6	9.8
	1984	486.1	2,215.8	720.2	3,422.1	10.5
Nuclear chemistry	1983	84.7	608.2	218.8	911.7	3.4
	1984	114.5	271.8	234.2	620.5	1.9
Prospecting, mining, and processing of nuclear materials	1983	580.7	857.0	238.4	1,676.0	6.3
	1984	698.4	432.5	253.2	1,384.1	4.2
Nuclear engineering and technology	1983	763.0	2,470.3	1,143.9	4,377.3	16.5
	1984	1,106.3	2,405.9	1,375.6	4,887.8	15.0
Application of isotopes and radiation in:						
Agriculture	1983	1,609.5	2,159.9	1,213.0	4,982.4	18.7
	1984	2,038.9	4,409.5	1,456.4	7,904.8	24.3
Medicine	1983	412.6	1,016.6	821.6	2,250.8	8.5
	1984	460.1	1,370.9	906.1	2,737.1	8.4
Biology	1983	20.0	111.1	171.4	302.5	1.1
	1984	31.0	38.2	87.3	156.5	0.5
Industry and Hydrology	1983	893.3	2,824.6	435.8	4,153.7	15.6
	1984	889.4	2,396.3	518.6	3,804.3	11.7
Safety in nuclear energy	1983	775.5	1,696.7	732.5	3,204.7	12.0
	1984	1,303.6	2,111.2	1,360.2	4,775.0	14.6
Miscellaneous[a]	1984	32.4	54.3	32.1	118.8	0.4
Total	1983	6,163.6	14,746.3	5,705.5	26,615.4	100.0
	1984	8,018.3	17,276.4	7,286.8	32,581.5	100.0

Source: IAEA, *Technical Cooperation Activities in 1984*. Report by the Director General GC(XXIX)/INF/226, August 1985.

[a]Miscellaneous amounts for 1983 were prorated by field and program component.

in the making of steel and paper and in the vulcanization of rubber. Nuclear engineering programs include assistance to member states in research reactor utilization, cobalt-60 irradiation facilities, and work on nuclear fusion. In hydrology, the agency, in cooperation with FAO and UNESCO, has applied isotope techniques to trace underground waters and to investigate water resources in arid regions of the world.

OTHER AGENCY INSTITUTIONS AND INTERESTS

In 1957, a report issued by the Preparatory Commission of the IAEA, stated: "At the start of its operations the only course open to the

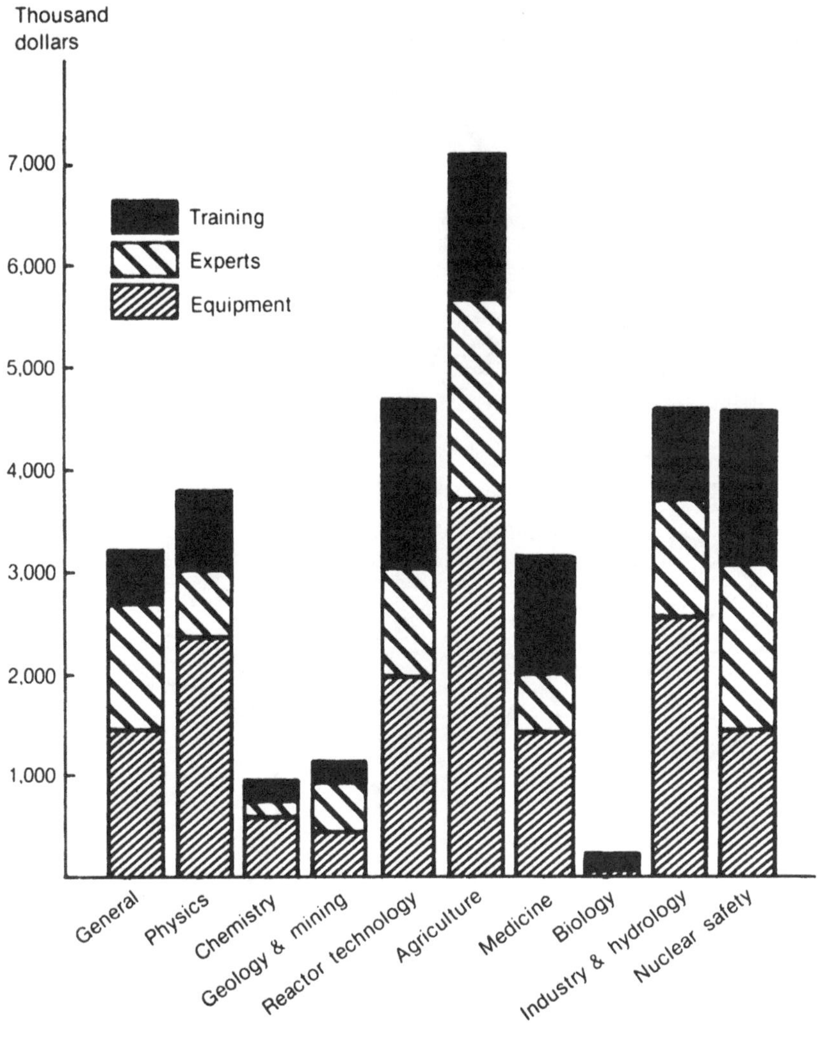

Figure 3-4. Utilization of resources, 1985. *Source*: IAEA, *The Agency's Technical Cooperation Activities in 1985*, Report by the Director-General, GC(XXX)/INF/234, August 1986, p. 52.

Agency will be to contract out all its laboratory work; but the nature of some of this work will make it desirable for the Agency to carry it out in its own facilities."[19] The Austrian Society for Atomic Energy Studies had a laboratory and a small research reactor at its center at Seibersdorf, about twenty miles southwest of Vienna. The agency decided to construct its own laboratory adjacent to the Austrian Reactor Center. Construction began on September 28, 1959, and the

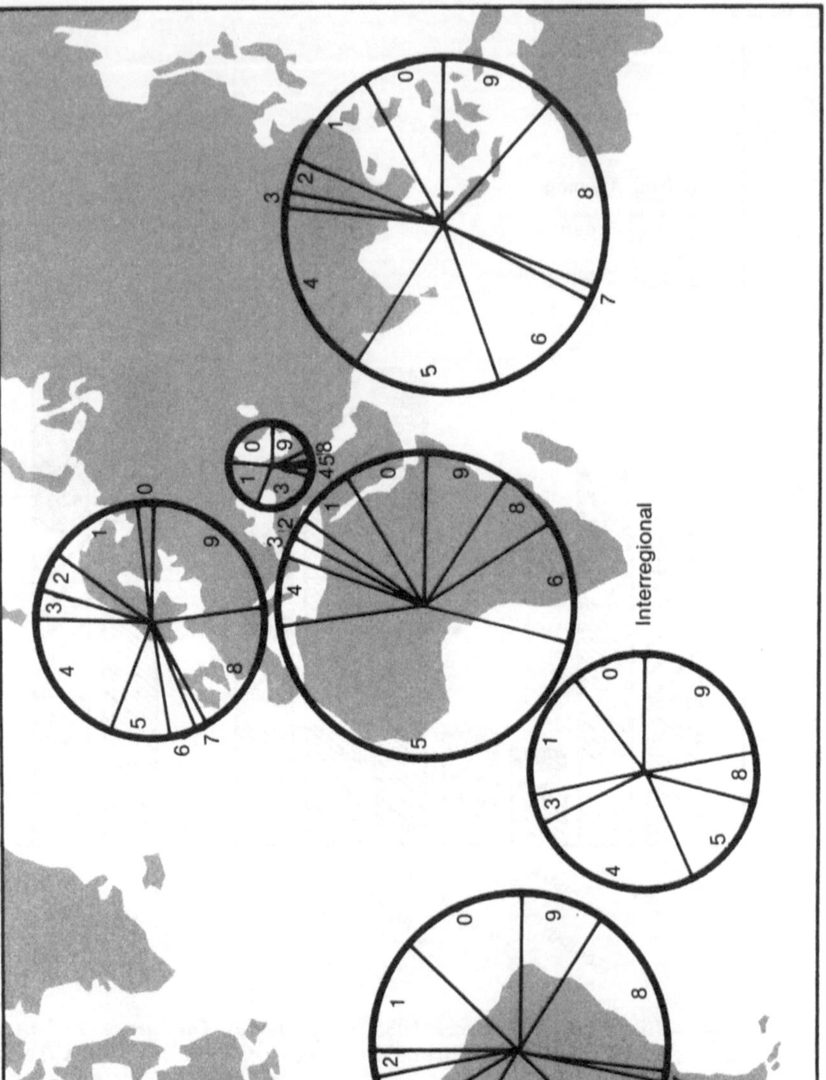
Interregional

SUMMARY
(thousands of dollars)

Field of activity	Africa	Asia and the Pacific	Europe	Latin America	Middle East	Inter-regional	All regions
0—General atomic energy development	656.7	799.7	87.6	1,063.5	123.5	487.7	3,218.7
1—Nuclear physics	433.5	1,026.5	614.5	853.5	103.1	778.0	3,809.1
2—Nuclear chemistry	158.9	292.0	227.8	263.3	—	—	942.0
3—Prospecting, mining and processing of nuclear materials	145.7	116.2	167.0	413.3	146.3	156.7	1,145.2
4—Nuclear engineering and technology	509.0	1,635.5	840.6	567.3	19.8.	1,138.6	4,710.8
Application of isotopes and radiation in:							
5—Agriculture	3,098.7	1,437.0	392.6	1,561.5	10.1	604.4	7,104.3
6—Medicine	893.0	1,135.4	177.0	664.5	—	309.0	3,178.9
7—Biology	7.7	108.5	53.3	53.5	—	—	223.0
8—Industry and Hydrology	456.9	1,840.1	852.5	1,435.1	28.0	12.4	4,625.0
9—Safety in nuclear energy	645.6	1,137.4	997.8	727.0	93.7	996.0	4,597.5
Subtotal	7,005.7	9,528.3	4,410.7	7,602.5	524.5	4,482.8	33,554.5
Miscellaneous	—	—	—	—	—	—	161.4
GRAND TOTAL	7,005.7	9,528.3	4,410.7	7,602.5	524.5	4,482.8	33,715.9

Figure 3-5. Distribution of technical cooperation inputs by field and region, 1985. *Source*: IAEA, *The Agency's Technical Co-operation Activities in 1985*, Report by the Director-General, GC(XXX)/INF/234, August 1986.

IAEA Seibersdorf laboratory was officially opened in September 1961. Much of the agency's in-house research and development in applied physics, chemistry, hydrology, agriculture, biology, and nuclear medicine is carried out at the Seibersdorf laboratory. The laboratory also provides services and does work for member countries that do not themselves have the necessary facilities to do the work. Under the agency's fellowship program, fellows and trainees from member countries come to Seibersdorf to work on specific projects or to obtain general training. The agency also maintains a Safeguards Analytical Laboratory at Seibersdorf.

Following the recommendations of the IAEA SAC in January 1961 and a subsequent exchange of views between the agency and the principality of Monaco on the impact of radioactivity on the marine environment, a tripartite agreement was concluded between the IAEA, Monaco, and the Oceanographic Institute—"Fondation Prince Albert Ier de Monaco."[20] Under this agreement, an International Institute of Marine Radioactivity was created in Monaco in 1961. The agency's Monaco laboratory, housed in a wing of the Oceanographic Museum, carries out studies of radioactivity in the marine environment and has a number of joint projects with the United Nations Environment Programme (UNEP). The objectives of the agreed research program, based on the recognition that "scientific knowledge about the redistribution of materials in the oceans, in particular by means of the biological cycle, is very limited," were defined in a board of governors document in February 1961 as:

1. The acquisition of knowledge concerning the movement of water and of marine organisms and the deposition of organic and inorganic matter.
2. The special study of the distribution and redistribution in marine organisms, ranging from phytoplankton to fish, of various forms of radioactive materials already existing in or that may be introduced into different locations.
3. The effects of radioactive materials at various concentration levels on the marine ecology.[21]

This laboratory, though comparatively small, continues to carry out useful research and training in marine science and particularly in environmental monitoring and in research on radioactive and nonradioactive pollutants in the sea.

Several developing countries have produced some outstanding theoretical physicists. Many of these countries, however, although eager to foster scientific research, have inadequate means to do so. As a result, many of their brightest young scientists leave the country to continue their research and studies elsewhere, often never to return. In 1960, Pakistan sponsored a resolution at the IAEA General Conference proposing the establishment of a theoretical research center where such scientists could meet and work and then return to their own countries. A panel of scientists, convened by the IAEA director general in 1961, strongly recommended the idea. A second scientific panel in 1963 drew up a detailed plan for such a center. Offers to host the center came from Pakistan, Italy, Denmark, and Turkey. In June 1963, the board of governors accepted the offer of the government of Italy to establish the center at Trieste for a four-year trial period.

The International Center for Theoretical Physics in Trieste, Italy, was formally inaugurated by the director general on October 5, 1964.[22] When the four-year term of the center's operation was over, the IAEA agreed to continue its support of the Trieste center and to increase its financial contribution. The government of Italy also agreed to continue its financial support. In 1970, UNESCO joined with the IAEA as an equal partner in support of the center, although the IAEA continued to retain responsibility of administering the center's operations. The center also receives support from other agencies and foundations such as the UNDP, the Ford Foundation, and the Swedish International Development Authority (SIDA). The Trieste center has been directed from its inception by the Pakistani theoretical physicist and Nobel laureate, Abdus Salam. The center serves both as a research facility and as a scientific training center. Its theoretical research program includes fundamental particle physics, plasma physics, solid-state physics, mathematics, biophysics, and geophysics.[23]

In addition to operating its laboratories at Seibersdorf and Monaco and the Trieste center the agency cooperates with and supports some regional research centers. One of the most important and active of such regional laboratories is the Middle East Regional Radioisotopes Center for Arab Countries in Cairo, Egypt, originally created in 1955 as a national Egyptian radioisotope research facility. At the request of the mid-east Arab countries, the IAEA board of governors, in June 1960, agreed to re-form it as the Middle Eastern Regional Radioisotopes Center for Arab Countries and the agreement came into force in January 1963.[24] The principal research programs of the Cairo center involve the use of radioisotope techniques in hydrology, agriculture, insect control, and medicine. These techniques are applied particularly to the solution of problems of the arid and semiarid regions. The center also trains Middle Eastern Arab scientists in the application of radioisotopes in science, agriculture, and industry.[25]

The Special Case of the Least Developed Member States

In its assistance to the Third World and developing countries, the IAEA has been making a special effort to aid the least developed countries. These have been defined by the United Nations General Assembly on the basis of economic criteria, such as very low annual per capita gross national product (under $350 in 1984), and social criteria, such as low literacy rates (under 20 percent). Almost two-thirds of the forty so-defined least developed countries are in Africa. Most of these countries are not members of the IAEA. Agency assistance to the twelve that are IAEA member countries is primarily in the areas of agriculture, medicine, and basic physics. Technical assistance given to these countries has risen significantly over the last four years. This is illustrated in table 3-4.

Technical Cooperation—Information and Safety

Information

Not all of the agency's programs involve technical assistance to the developing countries; some are of value to the technically advanced member countries as well. One such program is the International Nuclear Information System (INIS). Plans for a computer-based international nuclear information system were first discussed at the agency in 1968[26] and considered favorably at the general conference that year. Under the agency's Division of Scientific and Technical Information, INIS began operation in May 1970. Each participating member state agreed to transmit input for INIS from its scientific and technical literature. INIS was to correlate the data and make available to the participating member a magnetic tape service. INIS also issues publications and reports and provides microfiches to subscribers. Initially, the participating members were primarily the nuclear advanced countries. Now many developing countries have also found it useful to join INIS, thereby gaining access to information on developments in nuclear science throughout the world that might not otherwise be available to them. At present, eighty-eight IAEA member states and international organizations participate in INIS. The size of the data base exceeds one million records and has been expanding at the rate of 80,000 records a year. In 1984, the INIS

TABLE 3-4. IAEA ASSISTANCE TO LEAST DEVELOPED MEMBER STATES, 1981–1985
(thousands of dollars)

Member state	1981	1982	1983	1984	1985[a]
Afghanistan	38.7	33.6	157.2	28.3	14.0
Bangladesh	239.3	414.6	473.2	596.0	999.0
Ethiopia	24.1	23.8	63.4	110.4	197.2
Haiti	—	—	—	—	0.5
Mali	86.1	174.2	185.9	201.2	141.1
Nicaragua	—	—	—	13.2	92.3
Niger	41.7	35.5	145.0	152.8	175.3
Senegal	—	—	65.4	182.4	149.9
Sierra Leone	—	—	98.0	80.6	66.8
Sudan	227.3	206.2	176.2	297.2	491.8
Uganda	6.3	36.0	46.6	35.9	73.3
U.R. Tanzania	199.8	164.8	157.5	233.4	359.5
Total	863.3	1,088.7	1,568.4	1,931.4	2,760.7

Source: IAEA, *The Agency's Technical Cooperation Activities in 1984* GC(XXIX)/226 *The Agency's Technical Cooperation Activities in 1985* GC(XXX)INF/234.

[a]Figures for 1985 have been added to the original table.

Clearinghouse distributed 530,000 microfiches of publications, reports, and other documents to its subscribers.

Using INIS as a model, the FAO, in 1974, established an International Information System for the Agricultural Sciences and Technology (AGRIS). INIS performs the central processing for AGRIS.

Safety

As more and more of the member countries of the IAEA began to set up nuclear research facilities, to use radioisotopes in industry, agriculture, and medicine, and to build and operate nuclear power plants, agency emphasis began to focus not only on assistance in the provision and utilization of nuclear equipment and material, but also on safety-related matters. As a consequence, the agency's nuclear safety program has been growing continuously. In 1984, for example, almost 15 percent of the resources available for technical cooperation (including UNDP and extrabudgetary funds) was allocated for assistance in matters related to nuclear energy safety. This was exceeded only by the agency programs in agriculture (24 percent) and in nuclear engineering and technology (21 percent).

Nuclear safety is ultimately a national responsibility. Only the state has the authority to legislate and enforce. Nevertheless, in 1974 the IAEA inaugurated a Nuclear Safety Standards program (NUSS) to develop internationally agreed nuclear plant safety standards. Since then, NUSS has developed five codes of practice and fifty-five safety guides.[27] Although only recommendations, these codes and guides have increasingly been used as the basis for a nation's regulatory requirements, especially in the areas of reactor siting and quality assurance, and are slowly but steadily being incorporated into national laws. The Chernobyl accident has generated much discussion about the possibility for establishing binding international safety standards, possibly by converting NUSS codes and guides into mandatory provisions, but this is likely to require extensive and long-term discussions.

In 1981, in the aftermath of the Three Mile Island accident, a number of measures were taken to expand and intensify IAEA safety activities. An expanded emergency assistance program was developed and implemented. In 1982, a program of operational safety advisory review teams (OSARTs) was instituted to conduct comprehensive short-term on-site reviews of nuclear facilities. OSART visits depend upon an invitation from the state, and the proceedings are advisory and confidential, as is the resulting report. Importantly, OSART requests have come not only from developing countries, but also from advanced nuclear states such as France, Sweden, and the Netherlands. Following the Chernobyl accident, the number of requests for OSARTs has increased significantly.

The Incident Reporting System (IRS) was implemented in 1983. Based on the increasingly recognized importance of collating operational feedback experience in the area of reactor operation, IRS is intended as a mechanism for reporting and sharing information on safety-relevant incidents occurring in nuclear facilities.[28] The IRS is not an accident-reporting system and is not designed to deal with emergency situations. Rather, its objective is to share information in the interest of taking preventive measures against severe nuclear incidents. Participation in the IRS is voluntary, and although participating states are expected to report relevant information about safety-related incidents in nuclear power plants, they are not legally obligated to do so. Its effectiveness depends not only on close adherence to the commitment to report incidents, but also on the effectiveness of the national regulatory system that collects such information in the first instance. At present, twenty-four countries with nuclear power programs participate in IRS.

In 1984, two further initiatives to augment the agency's safety program were taken. One was the establishment of a program of radiation protection advisory teams (RAPATs) to assist states in assessing radiation protection needs and national regulatory arrangements in matters involving radiation. The RAPATs are also charged with developing coordinated long-term assistance plans in cooperation with WHO and the International Commission on Radiological Protection.[29]

The second initiative involved the establishment (by the director general) of an International Nuclear Safety Advisory Group (INSAG) to serve as a forum in which to examine nuclear safety issues and to provide advice. INSAG has already played a significant role in reviewing a special nuclear safety program proposed by the secretariat at the request of the board of governors meeting in special session in the wake of the Chernobyl accident.[30] INSAG also prepared an account of the accident, as well as recommendations for future action, based on a post-Chernobyl accident review held by the IAEA in August 1986 pursuant to a decision of the special board session.[31]

One initiative in support of enhanced international nuclear safety, however, failed. In 1981, the United States unsuccessfully proposed a convention establishing binding international rules on nuclear safety. Much earlier, in 1963, the Nordic countries had concluded a Mutual Emergency Assistance Agreement in connection with radiation accidents;[32] and in 1964, the general conference had requested the board of governors to take the necessary steps to stimulate conclusion of emergency agreements between two or more member states and the agency,[33] but governments were generally reluctant to develop binding rules on nuclear safety, mainly because of concerns that in doing so they would intrude on issues that lie within their domestic jurisdiction. The most that could be achieved was the establishment of guidelines for cooperation in response to major nuclear accidents[34] and for considering recommendations regarding transboundary radioactivity incidents.[35]

Chernobyl dramatically altered that situation. In July and August 1986, pursuant to a decision of the special board of governors meeting of May 21, 1986, two conventions were drafted to establish a representative group of government experts. These experts were to draw up international agreements on early notification of nuclear accidents and emergency response in the event of such an accident. The agreements were endorsed by the board of governors in September 1986, and submitted to the thirtieth general conference where they were approved by acclamation. Within hours of approval, fifty-one countries had signed the conventions.[36] Thus, a new chapter had opened in international nuclear safety, and the IAEA's agenda is now and for the foreseeable future heavily charged with important nuclear safety issues.

Physical Protection

Although the physical protection of nuclear material is a national responsibility, the IAEA, recognizing the importance of improving the protection of nuclear material both within a country and while in international transit, has played a major role in the development of a Convention on the Physical Protection of Nuclear Material.[37] The IAEA is the depository of instruments of ratification for the convention. The general conferences of 1983 and 1984 each passed resolutions urging its members to adhere to the convention; at present, eighteen states have ratified. An additional twenty-seven countries have signed the convention but not yet ratified it. The convention must be ratified by twenty-one states for it to enter into force. Table 3-5 lists the signatories to the convention as of August 1986.

Nuclear Power

Nuclear power can be an important adjunct to the energy programs of both the industrialized and developing countries. The IAEA cooperates with its member states in all aspects of their nuclear power programs. It helps developing countries determine their future electricity requirements and assesses the role that nuclear energy can play in satisfying those requirements. To this end, the IAEA sends planning missions and advisory teams to member countries requesting such assistance, and also has an effective training program, sponsoring seminars, giving training courses, publishing training manuals and guidebooks, and arranging for visits to facilities in countries with operating plants. For the majority of member countries, the conventional large nuclear power reactors, which are available from the principal power plant builders throughout the world, are far too large to be of value in addressing their energy needs economically. The IAEA has therefore played an active role in studying the application of small and medium nuclear power reactors in smaller or less industrialized countries. One such in-depth study, initiated by T. Keith Glennan, was carried out in the early 1970s.[38] In view of new and better designs for small and medium power reactors, which may now be applicable for both developing and industrialized countries, the IAEA has moved forward with a new study and an initial phase I report has been issued. A symposium on small and medium power reactors was held in conjunction with the September 1985 general conference, and a technical committee is reviewing the next steps to be taken.[39]

The agency also has carried out studies on the economics of nuclear power and its competitiveness with fossil-fired power plants under various conditions. As part of its program of support for nuclear power development in those member states where it makes economic sense, the IAEA has produced a number of useful reports and guidebooks. Reports are issued annually on *Power Reactors in the World*,

Performance Analysis, and *Operating Experience with Nuclear Power Reactors in Member States*. Guidebooks issued by the IAEA also deal with such matters as economic evaluation of bids for nuclear power plants, qualification and training of operations personnel, nuclear power plant control and instrumentation, and nuclear power project management, as well as safety issues involving plant siting, design, and operation. In addition to its activities related to proven nuclear power reactors, the agency reviews trends and developments in advanced reactor concepts. Meetings and studies have been organized on the prospects for gas-cooled reactors and an international working group (IWG) on fast breeders has held specialist meetings on liquid metal fast breeder reactors. A coordinated research program to assess the potential of advanced reactors, covering not only high temperature reactors and fast breeders but also advanced light water reactors and heavy water reactors, is also under consideration.

Hand-in-hand with its programs of assistance in nuclear power plant development and utilization, the agency cooperates with its member countries as well as with other international organizations in all areas of the nuclear fuel cycle—from studies on the availability of uranium-bearing ores through fuel fabrication and utilization to spent fuel management and waste disposal. Despite a slowing down of nuclear power reactor construction throughout the world, it is currently estimated that uranium requirements over the next decade will increase by a factor of two. The IAEA has been cooperating with the OECD's Nuclear Energy Agency (NEA) for a number of years through participation in a Joint NEA-IAEA Steering Committee and has participated in several joint activities in the uranium area. When the IAEA concluded that it could not continue its participation in these joint activities because of the inclusion of South Africa—a major uranium supplier and hence a key participant in the joint programs, it so advised the NEA in June 1984. The programs affected were the Joint Steering Group on Uranium Resources, the Joint Working Group on Uranium Extraction, the Joint Working Group on Uranium Resources, and the Joint Group of Experts on Research and Development in Uranium Exploration Techniques. However, in view of the importance of the work, an effective arrangement was worked out. It was agreed that while the IAEA would no longer participate in the joint working groups with the NEA in the uranium area, its secretariat will continue to contribute to future editions of the uranium "red book" (*Uranium Resources, Production and Demand*) calling on specific experts on an ad hoc basis as required. Thus, in place of the former joint NEA-IAEA uranium working groups, the NEA will set up its own uranium group with which the IAEA Secretariat can

TABLE 3-5. STATUS OF SIGNATURES AND RATIFICATIONS AT THE CONVENTION ON THE PHYSICAL PROTECTION OF NUCLEAR MATERIAL

Name of state/organization	Date of signing	Place of signing	Ratified
1. United States	3 March 1980	New York, Vienna	13 December 1982
2. Austria	3 March 1980	Vienna	
3. Greece	3 March 1980	Vienna	
4. Dominican Republic	3 March 1980	New York	
5. Guatemala	12 March 1980	Vienna	23 April 1985
6. Panama	18 March 1980	Vienna	
7. Haiti	9 April 1980	New York	
8. Philippines	19 May 1980	Vienna	22 September 1981
9. German Democratic Rep.	21 May 1980	Vienna	5 February 1981
10. Paraguay	21 May 1980	New York	6 February 1985
11. USSR	22 May 1980	Vienna	25 May 1980
12. Italy[a]	13 June 1980	Vienna	
13. Luxembourg[a]	13 June 1980	Vienna	
14. Netherlands[a]	13 June 1980	Vienna	
15. United Kingdom[a]	13 June 1980	Vienna	
16. Belgium[a]	13 June 1980	Vienna	
17. Denmark[a]	13 June 1980	Vienna	
18. Fed. Rep. of Germany[a]	13 June 1980	Vienna	
19. France[a]	13 June 1980	Vienna	
20. Ireland[a]	13 June 1980	Vienna	
21. EURATOM	13 June 1980	Vienna	
22. Hungary	17 June 1980	Vienna	4 May 1984
23. Sweden	2 July 1980	Vienna	1 August 1980
24. Yugoslavia	15 July 1980	Vienna	14 May 1986
25. Morocco	25 July 1980	New York	
26. Poland	6 August 1980	Vienna	5 October 1983
27. Canada	23 September 1980	Vienna	21 March 1986
28. Romania	15 January 1981	Vienna	
29. Brazil	15 May 1981	Vienna	17 October 1985
30. South Africa	18 May 1981	Vienna	

31. Bulgaria	23 June 1981	Vienna	10 April 1984
32. Finland	25 June 1981	Vienna	
33. Czechoslovakia	14 September 1981	Vienna	23 April 1982
34. Republic of Korea	29 December 1981	Vienna	7 April 1982
35. Norway	26 January 1983	Vienna	15 August 1985
36. Israel	17 June 1983	Vienna	
37. Turkey	23 August 1983	Vienna	27 February 1985
38. Australia	22 February 1984	Vienna	
39. Portugal	19 September 1984	Vienna	
40. Niger	7 January 1985	Vienna	
41. Liechtenstein	13 January 1986	Vienna	
42. Mongolia	23 January 1986	New York	28 May 1986
43. Argentina	28 February 1986	Vienna	
44. Spain[a]	7 April 1986	Vienna	
45. Ecuador	26 June 1986	New York	
46. Indonesia	3 July 1986	Vienna	

Note: Opened for signature at Vienna and New York March 3, 1980.
Source: IAEA, GC(XXX)/INF/236 Annex 26, 1986.
[a]Signed as EURATOM state.

cooperate and the IAEA has agreed, for its part, to assume major responsibility for uranium exploration and ore-processing technology reviews, to which the NEA will be able to contribute. The necessary input from South Africa would continue to be obtained, although the IAEA will not have to deal with South Africa directly. The IAEA continues to assist its member countries in this area by supporting visits of technical experts and by holding interregional training courses in uranium exploration, uranium recovery from nonconventional resources, and ore processing.

In the area of fuel fabrication and fuel technology, the agency has organized specialist meetings and seminars on water reactor fuel technology and on fuel element cladding interactions with water coolant in power reactors. This has been done under the aegis of the International Working Group on Water Reactor Fuel Performance and Technology. The agency has established a Nuclear Fuel Cycle Information System that collects information on the status of nuclear fuel cycle facilities and contains data on plants and facilities for ore processing, refining and conversion, uranium enrichment, fuel fabrication, spent fuel storage, and fuel reprocessing.

As power reactor operators accumulate more and more spent fuel, its storage and management becomes a problem requiring careful attention. The IAEA has addressed this problem by holding seminars on the technical and environmental aspects of spent fuel management and by collecting information on fuel storage options and on spent fuel transport, reprocessing, and recycling technology. A coordinated research program on the behavior of spent fuel assemblies under long-term storage is being planned as a future agency activity. An IAEA conference on radioactive waste management, held in Seattle in 1983, covered the entire range of radioactive waste management including technological, environmental, regulatory, and policy issues. A number of agency guidelines and technical reports have been issued on these subjects, and a draft code of practice on the management of radioactive waste from nuclear power plants has been circulated. Research on evaluation of high-level waste solidification and related technologies has also been carried out under agency programs.

Committee on Assured Supply

In conjunction with its program of assistance and support of the nuclear power programs of its member states, the IAEA is addressing the problem of assurance for these countries of adequate supply, under appropriate nonproliferation conditions, of material and components necessary for their nuclear power programs. To this end, a

Committee on Assurances of Supply (CAS) was created in 1980 with the objective of establishing procedures in international nuclear commerce and cooperation that would reduce uncertainties in nuclear supply without compromising nonproliferation objectives. A number of CAS working groups have been formed, each addressing different aspects of supply assurances.[40] One CAS working group has been seeking to develop a set of draft principles, involving such contentious matters as the interrelationship between supply assurances and nonproliferation guarantees, adherence to the nonproliferation treaty, the role of safeguards in verifying nonproliferation undertakings, and the issue of attacks on peaceful nuclear facilities. Another working group on an IAEA system for an emergency and back-up mechanism has proposed a scheme under which the IAEA would serve as an information clearinghouse for materials that may be made available by suppliers in case of a supply emergency; the IAEA would also determine the conditions under which the materials could be obtained. A third CAS working group has been studying mechanisms for revising international nuclear cooperation agreements and has produced a set of agreed understandings on revision mechanisms. CAS meets in plenary session several times a year and has, through the end of 1985, held seventeen sessions.

Nuclear Safeguards

Little has been said thus far about the important role the IAEA plays in nuclear safeguards, since that will be the subject of much of the remainder of this book. It is appropriate, however, to discuss at this point the agency's safeguards organization and responsibilities. The Department of Safeguards, formerly headed by an inspector general for safeguards and now directed by an IAEA deputy director general, has three operations divisions responsible for carrying out safeguards inspections, and four non-operations divisions responsible for development and technical support, data processing and information treatment, evaluation, and training and administration (see figure 3-3). Most of the members of the staff of the operations divisions are professionals and many serve as inspectors. They are assisted by a general staff, which includes inspection assistants. Although the non-operations divisions have a smaller fraction of professional staff members, many of them are qualified as inspectors and do on occasion work in the operations divisions as inspectors. The safeguard directorate currently has a staff of 455, of whom 272 are in the professional category; 186 serve as inspectors. The safeguards staff has grown significantly since its original 1960 staff of seven, four of whom were professionals. The number of safeguards inspections, and hence the size of the safeguards staff, increased markedly after the approval of INFCIRC/66 in 1965, its revision in 1968, and especially after the adoption of INFCIRC/153, which implemented the nonproliferation treaty's safeguards commitment. The increase in safeguards staff of the agency was accompanied by a corresponding increase in U.S. nationals joining the safeguards staff. Figures 3-6 and 3-7 show the growth of the Department of Safeguards (DSG) between 1960 and 1985. There are currently 50 U.S. nationals on the professional staff of the DSG, including a number of cost-free experts. This corresponds to almost 19 percent of the professional staff. About 10 percent of the professionals on the agency's safeguards staff are U.S. cost-free experts.

Safeguards are by far the largest single item in the IAEA budget, accounting for more than one-third of its appropriations. Of the $96.83 million appropriations for the IAEA regular budget for the year ending December 31, 1984, $32.56 million was allocated for safeguards. For 1986, the regular budget is $98.68 million, $33.34 million of which is allocated to safeguards—$20.46 million for safeguards implementation and $12.88 million for safeguards development and support. An additional $3.33 million is available for safeguards development and support from extrabudgetary resources

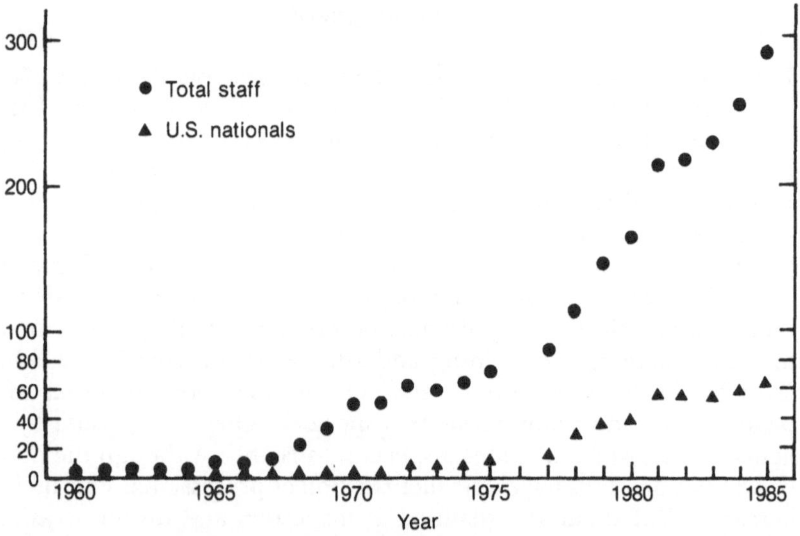

Figure 3-6. Professional staff in Department of Safeguards. *Source*: Kenneth E. Sanders, "International Safeguard Professionals," Proceedings of the 27th Annual Meeting of the Institute of Nuclear Materials Management, New Orleans, La., June 22–25, 1986.

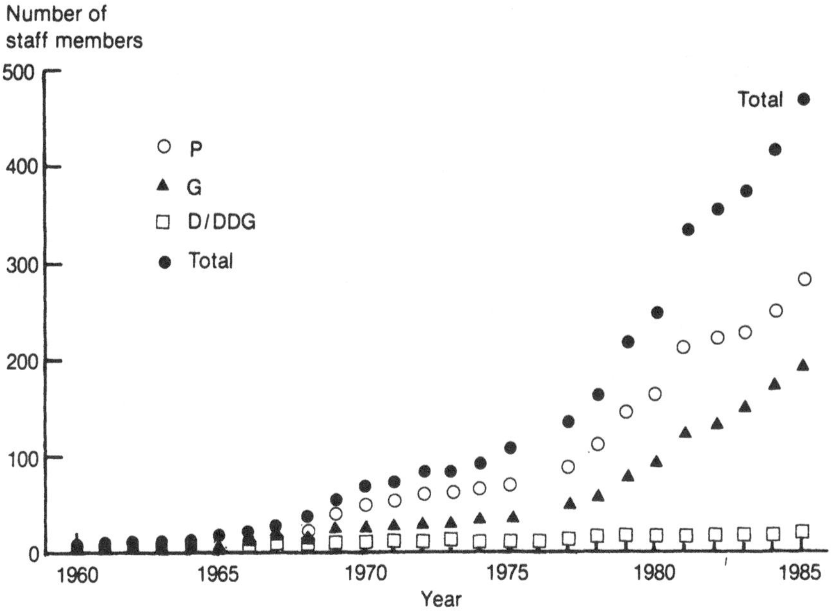

Figure 3-7. Department of Safeguards Growth, 1960–1985. *Source*: Kenneth E. Sanders, "International Safeguard Professionals," Proceedings of the 27th Annual Meeting of the Institute of Nuclear Materials Management, New Orleans, La., June 22–25, 1986.

(which come from voluntary contributions). The United States accounts for two-thirds of this extrabudgetary contribution.

Because of the stringent educational and experience requirements for employment in the DSG, a majority of the safeguards staff comes from the industrialized countries (figures 3-8 and 3-9). To qualify as a professional inspector, one must have a university degree in the physical sciences or engineering, significant experience in the nuclear field, and fluency in English, French, Russian, or Spanish.

When the agency's safeguards responsibilities and staffing requirements underwent rapid expansion following the application of INFCIRC/66 and INFCIRC/153, it became necessary to translate the safeguards objectives into implementation. There was an increasing trend toward raising safeguards problems and issues in the board of governors. This resulted in a tendency to treat such issues from the national point of view of the member states on the board. As an alternative, a Standing Advisory Group on Safeguards Implementation (SAGSI) was created. SAGSI serves as an advisory group to the director general. Its members, like those of the Scientific Advisory

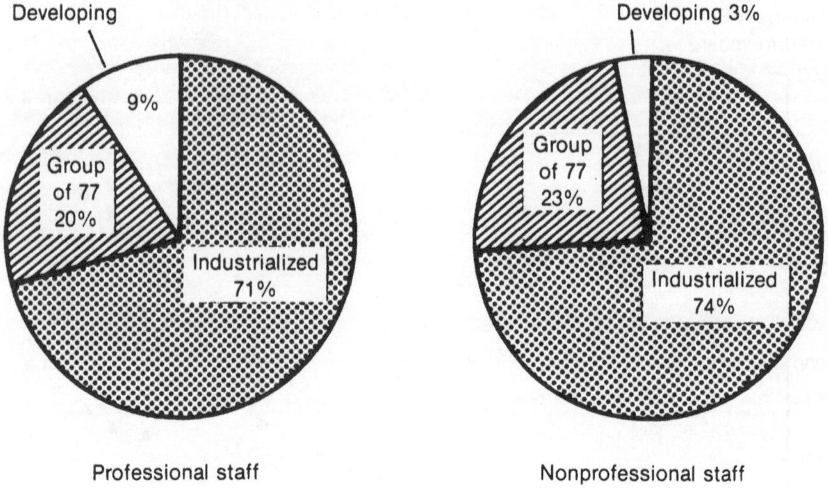

Figure 3-8. Distribution of the Department of Safeguards staff by group, 1985. *Source*: Kenneth E. Sanders, "International Safeguard Professionals," Proceedings of the 27th Annual Meeting of the Institute of Nuclear Materials Management, New Orleans, La., June 22–25, 1986.

Committee, nominally participate as independent experts rather than as representatives of their countries, although this is a separation that is sometimes difficult to make.

NOTES

1. For a comprehensive review and analysis of the structural, functional, and operational features of the agency during its first 15 years, see Lawrence Scheinman, "IAEA: Atomic Condominium?" in Robert W. Cox and Harold K. Jacobson, eds., *The Anatomy of Influence: Decision Making in International Organization* (New Haven, Conn., Yale University Press, 1973) chapter 7.

2. U.S. Congress, United States Senate, Hearings on the IAEA Statute, 85 Cong. 1 sess. March 22, 1957.

3. The first board of governors was made up of representatives from Argentina, Australia, Brazil, Canada, Czechoslovakia, Egypt, France, Guatemala, India, Indonesia, Italy, Japan, Korea, Pakistan, Peru, Portugal, Romania, Sweden, Turkey, South Africa, USSR, UK, and the United States. IAEA document GOV/INF/4 (Vienna, IAEA, October 10, 1957).

4. The People's Republic of China also could have been named as the regionally most advanced state, but that would have affected India, who vigorously opposed formulations altering its status.

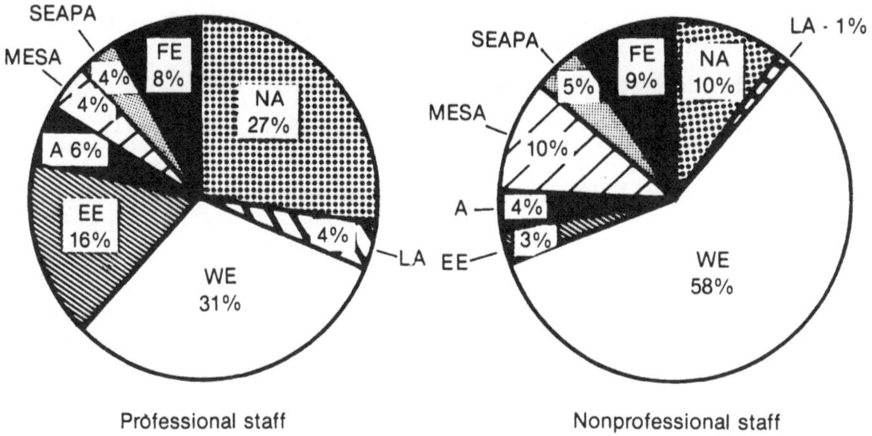

Geographical Region:

NA — North America

LA — Latin America

WE — Western Europe

EE — Eastern Europe

A — Africa

MESA — Middle East and South Asia

SEAPA — South East Asia, Pacific, and Australia

FE — Far East

Figure 3-9. Distribution of Department of Safeguards staff by geographical region. *Source*: Kenneth E. Sanders, "International Safeguard Professionals," Proceedings of the 27th Annual Meeting of the Institute of Nuclear Materials Management, New Orleans, La., June 22–25, 1986.

---

5. IAEA Statute, Article V. F.2.

6. An example of proposing an activity can be found in the General Conference Resolution GC(XXVI)/RES/399 (1982) calling upon the board of governors, in consultation with the director general, to conduct a review of agency activities over its first twenty-five years, resulting in IAEA, *Review of the Agency's Activities* GC(XVIII)/718 (Vienna, July 1984). An example of taking decisions on matters referred to it by the board of governors can be found in general conference adoption of the Convention on Assistance in the Event of Nuclear Accidents and Radiological Emergency and in Convention on Early Notification of a Nuclear Accident, both of which are contained

in the IAEA document GC(SPL.I)/2 Annex III and Annex II respectively (Vienna, IAEA, September 24, 1986).

7. IAEA document GOV/INF/22 (Vienna, IAEA, September 22, 1958).

8. IAEA document GOV/INF/33 (Vienna, IAEA, September 2, 1959).

9. Separate arrangements for financing safeguards activities were first established by GC(XIV)/RES/283 of 1971 and with some modifications of a minor technical nature have continued to the present time. The effect of a special formula for determining the allocation of safeguards costs results in seventy-six member states being shielded from increased safeguards costs and having their contributions to safeguards frozen at their level of contribution for 1976. This effectively means that these seventy-six states bear approximately 2.5 percent of total safeguards costs, while the remaining thirty-six unshielded member states underwrite the remaining 97 percent of costs, and contribute toward safeguards expenses on an increased scale. In 1983, a literal application of the formula worked out to shield certain member states from increased safeguards costs would have resulted in four member states that normally were not shielded being granted relief in respect of contributions to safeguards costs (Czechoslovakia, Ireland, Byelorussian SSR and Ukrainian SSR). To avoid this result, the general conference passed a resolution GC(XXVII)/RES/416 (1983), thereby maintaining the list of non-shielded members as it had been in previous years. However, since 1982, the general conference has been urging the board of governors to review financing arrangements for safeguards with a view to arriving at a long-term formula. Several different formulas have been proposed, but none has succeeded in drawing the necessary consensus for adoption. One of those proposals has introduced a new and complicating dimension into the discussions—it recommends that the nuclear weapon states, particularly the two superpowers, shoulder the main burden for financing safeguards.

10. Miscellaneous income includes income for work done on behalf of others (such as data processing, library or printing services), investment and interest income and gains resulting from currency exchange, and income derived from specific programs (such as laboratory income or program support costs).

11. In the wake of the Chernobyl accident, the board of governors at a special session in May 1986 requested the director general to present suggestions on implementation of a special program on nuclear safety. This included a post-accident review meeting (held in August 1986), the drafting of international conventions on early notification and coordinated emergency response in the event of nuclear accidents (negotiated over a four-week period from late July to mid-August and approved by the general conference in Special Session in September 1986), the holding of a special general conference on nuclear safety issues, and expansion of agency nuclear safety activities (in a variety of areas including technical and economic performance of nuclear power, radioactive waste management, human health, radiation protection, and safety analysis, among others). The board of governors approved proposals for approximately $2 million in expanded nuclear safety

activities for 1987. Extraordinary costs for 1986, as noted in the text above, were covered by reallocations from the regular budget and voluntary contributions that totalled $450,000 by October 1, 1986.

12. IAEA document GC(XXX)/INF/234 (Vienna, IAEA, 1986).

13. See IAEA *Bulletin*, vol. I no. 2 (1959) pp. 13–14.

14. IAEA *Bulletin*, vol. XV no. 1 (1973) pp. 31–34.

15. Ibid.

16. Ibid.

17. The current scientific advisory committee, whose term extends from September 1985 to September 1988, includes: D. Beninson (Argentina), D. Berenyi (Hungary), H. Bohm (Federal Republic of Germany), Z.Y. Chen (People's Republic of China), F. Culler (United States), J. Dunster (UK), A.A.R. El Agib (Sudan), G. Fernandez de la Garza (Mexico), L. Gutierrez-Jodra (Spain), J. Jennekens (Canada), M.wa Kalenga (Zaire), A. Ohyama (Japan), R. Ramanna (India), M.F. Troyanov (USSR), I. Ursu (Romania), and G. Vendryes (France). Eleven of these are reappointments; four are new.

18. INIS, a worldwide information system in nuclear science and technology, is dealt with later in this chapter.

19. IAEA *Bulletin*, vol. I no. 2 (1959) pp. 21–22.

20. See IAEA document GOV/INF/48 (February 14, 1961) Scientific Collaboration with the Government of Monaco. The activities at Monaco are discussed in "Radioactivity in the Seas," IAEA, *Bulletin*, vol. 8 no. 3 (1966) pp. 17–23.

21. Ibid.

22. IAEA *Bulletin*, vol. VII no. 1 (1965) pp. 31–33; vol. VII no. 3 (1965) pp. 3–10.

23. For a recent and comprehensive overview of the Trieste Center see, IAEA/UNESCO, "The International Centre for Theoretical Physics at Miramar-Trieste, 1964–1984" (IAEA/PI/A.8E) (Vienna, IAEA, 1984).

24. IAEA *Bulletin*, vol. V no. 3 (1963) p. 24.

25. IAEA *Bulletin*, vol. VIII no. 1 (1966) pp. 22–24. For a discussion of the politics underlying several of the activities discussed herein (Trieste, Cairo Center) see Scheinman (note 1 above).

26. IAEA *Bulletin*, vol. X no. 6 (1968) pp. 14–22. See also, I.S. Zheludev and H.W. Groenewegen, "INIS: The International Nuclear Information System," IAEA *Bulletin* vol. XX no. 4 (1978) pp. 7–17; and I.S. Zheludev and A.G. Romanenko, "Direct Access to INIS," IAEA *Bulletin* vol. 23 no. 1 (1981) pp. 11–14. A very recent review is IAEA, *INIS: The International Nuclear Information System* (Vienna, IAEA, March 1986) IAEA, PI/A.

27. There is one code of practice in each of the five areas in which the agency is involved in matters related to nuclear safety: government organization, siting, design, operation, and quality assurance. See also IAEA, *The Nuclear Power Safety Programme of the IAEA* (Vienna, IAEA Information Series, 1981).

28. The Incident Reporting System is described in IAEA, *The IAEA Incident Reporting System, IAEA-IRS* (IAEA/PI/B.26.E) (Vienna, IAEA Information Series, 1986).

29. The United States has been giving consideration to creating a Nuclear Safety Training Academy that, in cooperation with the IAEA, would give nuclear safety training to visiting scientists and engineers. A pilot course was offered at Stanford University in September 1985, drawing participants from Europe, Latin America, China, Korea, and the Philippines.

30. The proposed program was adopted by the board of governors at a special session in December 1986. This has become the basis for agency safety activities over a several-year period.

31. See INSAG, Summary Report on the Post-Accident Review Meeting on the Chernobyl Accident (Vienna, August 30–September 5, 1986) (IAEA document GC(SPL.I)/3) (Vienna, IAEA, September 24, 1986). The special session of the board requesting the post-accident review was discussed above.

32. See IAEA document INFCIRC/49 Nordic Mutual Emergency Assistance Agreement in Connection with Radiation Accidents (Vienna, IAEA, November 8, 1963).

33. See GC(VIII)/290 (Annex I) (1964), Draft Resolution on the Transport of Radioactive Materials.

34. IAEA document INFCIRC/310, Guidelines for Mutual Emergency Assistance Arrangements in Connection with a Nuclear Accident or Radiological Emergency (Vienna, IAEA, January 1984).

35. IAEA document INFCIRC/321, Guidelines on Reportable Events, Integrated Planning and Information Exchange in a Transboundary Release of Radioactive Materials (Vienna, IAEA, January 1985).

36. The two conventions were negotiated in record time. One was completed in two weeks, the other in four. The drafters had the benefit of INFCIRC/310 and INFCIRC/321 noted above and upon which they drew substantially. Of significance in the notification convention is the fact that the weapons states agreed that notification would extend to accidents arising in military facilities, although it is stated in the permissive "may" rather than the obligatory "shall." Article 1 of the notification convention states that the convention "shall apply in the event of any accident . . ." and Article 2 designates the facilities and activities covered by the Article 1 obligation. Article 3 speaks of other nuclear accidents and was drafted to cover military facilities. It states: "With a view to minimizing the radiological consequences, States Parties may notify in the event of nuclear accidents other than those specified in Article 1."

37. IAEA document INFCIRC/225 (corrected) The Physical Protection of Nuclear Material (Vienna, IAEA, February 1976).

38. IAEA, *Market Survey for Nuclear Power in Developing Countries* (Vienna, IAEA, September 1973).

39. Two barriers to increased use of nuclear power that have been identified are the absence of infrastructure requirements in many states (trained technicians, relevant support and service industries, adequate size electricity grids to accommodate anything other than a small reactor, etc.) and the unavailability of capital for investment in nuclear power development. The latter is a major stumbling block.

40. On the Committee on Assured Supply, see Charles N. Van Doren, *Nuclear Supply and Non-Proliferation: The IAEA Committee on Assurances of Supply*, A report prepared under a research contract for the Congressional Research Service, Report No. 83-202 S (CRS, October 1983).

Chapter 4

THE AGENCY'S SAFEGUARDS SYSTEM BEFORE THE NPT

Clearly, as we have seen in the preceding chapter, the substantial activities of the International Atomic Energy Agency (IAEA) dedicated to enlarging "the contribution of atomic energy to peace, health and prosperity throughout the world" are vitally important to the great majority of the agency's members.[1] But what makes the agency truly significant in world affairs is its safeguards responsibility. As noted in chapter 1, safeguards are both unique and evolutionary. They are unique in that they involve voluntary acceptance by sovereign states of verification (including on-site inspection within their territory by international officials) of compliance with their international nuclear undertakings and commitments. Heretofore, states were taken at their word or not without independent, objective verification of their activities with respect to whatever matter was under obligation. (Vanquished states, compelled by peace treaties to accept certain obligations, sometimes were subjected to verification; we are speaking here of voluntary undertakings by sovereign states limiting their freedom of behavior.) In contrast, international safeguards, as described by a former senior official of the IAEA are "the first significant attempt of our restless species to combine agreements on arms control with objective and effective verification . . . of . . . compliance. . . ."[2]

The scope of international safeguards, and the *relative* smoothness of their implementation (once a role and purpose for safeguards had been defined and the rules and procedures established) are all the more remarkable in light of the novelty of the activity involved. Provision for such verification, as we shall have occasion to note again, is not a reflection of suppliers' lack of confidence in the integrity of their nuclear trading partners. Rather, it attests to a recognition of the special nature of nuclear energy and of the importance of assuring third parties that their own security is not being jeopardized by the nuclear energy development activities of others. This perspective on safeguards has come to be increasingly widely shared. The existence, in particular, of comprehensive (that is, full-scope) international safeguards in regions where there is active nuclear energy development undoubtedly helps to avoid destabilization that otherwise might result from such activities.

EVOLUTION OF SAFEGUARDS CONCEPTS

The concept of safeguards itself changed substantially between the initial postwar period and the Atoms for Peace era initiated by President Eisenhower's speech at the United Nations in December 1953. At first, safeguards were conceived in very sweeping terms as an integral part of a system of international ownership, management, and control of sensitive nuclear facilities (that is, direct control of the means of weapons production)—the Baruch Plan—a system that never materialized.[3] The United Nations Atomic Energy Commission (UNAEC), created in 1946 to make proposals for controlling atomic energy in order to foster its use for peaceful purposes only, identified three kinds of abuse against which it would be necessary to protect—diversion of materials, clandestine operation, and seizure of material or facilities. Its definition of safeguards embraced virtually all forms of control ranging from material accounting and inspection, to licensing, to direct power and authority over decisions governing day-to-day operations, up to and including, where appropriate, direct international ownership.

Seizure was treated as a fundamentally political issue beyond the ambit of even such broadly conceived safeguards. As to diversion, the UNAEC concluded that while a system of inspection, including unrestricted access to all equipment and operations, would probably suffice for dealing with the initial stages of the nuclear fuel cycle (mining, milling, refining), management in varying degrees by the proposed international control agency would be necessary in the case of reactors and enrichment and reprocessing facilities.

With respect to clandestine activities, the UNAEC determined that in addition to direct managerial control over many nuclear operations, the international control agency would also require authority to seek out and detect such activities. The UNAEC emphasized the vital importance of preventing the unauthorized accumulation of fissionable material because it was felt that detecting clandestine activities would be much more difficult at later stages (such as actual bomb fabrication). The safeguards rights to these ends would have to include "broad privileges of movement and inspection, including the rights to conduct surveys by ground and air."[4] This far-reaching conception of safeguards is given attention here not because it represents a universal truth about what safeguards ought to be in order to have any value, but because some contemporary critics of IAEA safeguards, in emphasizing the importance of preventing proliferation, appear to be guided in their assessment of safeguards effectiveness by the stringent spirit of these early efforts to define international safeguards, and their approach tends to guide public perceptions of the purpose of safeguards.

Present-day safeguards are based on a fundamentally different approach that was developed as a result of the change in policy on nuclear cooperation, reflected philosophically in the Atoms for Peace speech, and practically in the revision of the Atomic Energy Act of 1946 to permit U.S. international cooperation with others.[5] The new approach abandoned the earlier notion of an international monopoly in favor of a system of international verification of obligations accepted by states with respect to *nationally* owned and controlled nuclear activities. An important underlying assumption of this approach is that despite the Janus-like character of nuclear technology, it is possible to achieve a separation between peaceful and military nuclear activities if appropriate assurances (in the form of national undertakings and pledges) are established and maintained. Such assurances are subject to an effective verification system based on international safeguards, including inspection.

Safeguards in this framework have a more limited yet even more central role in nonproliferation and the peaceful development of nuclear energy than in the earlier context just discussed. They are more limited in terms of their mandate and authority. They are not an arm of a supranational enterprise exercising world government responsibilities, as they would have been if the Baruch Plan and UNAEC report had been translated into reality. Rather, they are a specific means of verification for a more particular application, one that is defined by undertakings that states enter into and in support of which safeguards have a verification function. Safeguards may be comprehensive or limited to specific materials and facilities depending upon the scope of the underlying commitments to which they apply. Safeguards are not intended to seek out clandestine operations or undeclared activities or to govern or regulate national action.

Their function is to monitor, audit, and report; to *verify* that states are in compliance with their voluntary undertakings not to use nuclear materials and facilities under safeguards for proscribed (military or explosive) purposes.[6] Nor are they intended to prevent national accumulation of safeguarded weapons-usable material. Insofar as the latter is a nonproliferation policy objective, it has to be sought through other related means.[7]

Safeguards are more important because many of the institutions and instrumentalities that collectively constitute the nonproliferation regime (discussed in chapter 1) rely on their effects. The Non-Proliferation Treaty, the Nuclear Suppliers Guidelines, and the nuclear cooperation policies of the principal nuclear suppliers are dependent on effective international safeguards for their own utility in supporting nuclear nonproliferation and international cooperation and trade in nuclear materials, facilities, and technology. Thus, safeguards have become the crux of an effective international nuclear regime.

But if they are absolutely necessary, they are not alone sufficient. Safeguards need the reciprocal support of those other nonproliferation institutions and instrumentalities. Changing circumstances and conditions arising from developments in the nuclear field or from the international environment may give rise to still other mechanisms or institutions to complement and reinforce safeguards in the future. Clearly, if these mechanisms and safeguards are to effectively serve the cause of nonproliferation and peaceful international nuclear cooperation, a similar reciprocal relationship must exist between them.

The IAEA Statute and Safeguards

Negotiation of the IAEA statute and its safeguards provisions were discussed in chapter 2. Our concern here is with the statutory provisions themselves. The basic authority for the application of agency safeguards is its statute. The statute, it will be recalled, does not require IAEA members to submit to safeguards unless the member has sought and received assistance in some peaceful nuclear activity from the agency; neither does it oblige member states to condition their international nuclear assistance on the recipient's acceptance of such safeguards.[8]

Legal obligations to invoke safeguards are rather to be found in other treaty instruments such as the NPT and the Treaty of Tlatelolco; in bilateral agreements for cooperation consummated between nuclear suppliers and their recipients, and in multilateral understandings such as the Nuclear Supplier Guidelines (insofar as suppliers commit themselves to require IAEA safeguards on transferred materials as a condition for supply). For the safeguards obligations incurred, however, the IAEA statute establishes a framework for implementation.

Article II establishes as one of the two main purposes of the agency to "ensure, so far as it is able, that assistance provided by it or at its request or under its supervision or control is not used in such a way as to further any military purpose." Article III.A.5 authorizes the agency to "establish and administer safeguards" to this end. The same article also specifies the circumstances in which, and by whom, IAEA safeguards may be invoked. These include: circumstances where the agency itself is the source or channel of assistance; where the parties to a bilateral or multilateral arrangement request agency safeguards to be applied; and where a state unilaterally submits itself to agency safeguards. In practice, before the NPT took effect, most safeguards arrangements came under the second category and resulted from transfers of bilateral safeguards arrangements to the agency. Since 1970, however, IAEA safeguards responsibilities have resulted primarily from non-nuclear weapon states joining the NPT and thus becoming obliged to accept IAEA safeguards on all of their peaceful nuclear activities.

Article XII sets forth the rights and responsibilities of the agency with respect to any situation where it is authorized to apply safeguards. These include the right to review the design of specialized equipment and facilities, including reactors, to assure that the design "will permit effective application of . . . safeguards"; the right to "require the maintenance and production of operating records" concerning the utilization of nuclear material; and the right to "call for and receive progress reports." Most significant, and certainly the right that has earned the greatest notoriety, is the IAEA's authority to send inspectors into safeguarded states and granting them "access at all times to all places and data . . . as necessary to account for . . . materials . . . and to determine whether there is compliance with the undertaking against use in furtherance of any military purpose. . . ."[9] These provisions not only establish the agency's right of verification, but oblige safeguarded states to adopt practices facilitating the verification process.

It is essential to understand at the outset that in speaking of these rights, we are referring to a legal authorization for the agency to engage in and conduct such activities, not to a discretionary authority to impose them unilaterally. The actual application of safeguards is conducted on the basis of safeguards agreements negotiated between the agency and the safeguarded state, within the framework of the statute and agency safeguards documents that have been developed by the IAEA members and secretariat and approved by the board of governors and the general conference. As we shall see, these documents have put the basic concepts of the statute into operational form and have in some cases narrowed the agency's authority, although always consistent with the principle of ensuring that the agency can carry out its safeguards responsibilities. Even the statute itself did not establish unlimited authority. Reciprocal responsibilities and limitations reflecting the sensitive nature of the proscribed activities were also stipulated in the statute. Thus, the purposes of these rights are narrowly defined to apply only "to the extent relevant to the project or arrangement."[10] Design-approval authority exists only for the purpose of "assuring that (the facility in question) will not further any military purpose."[11] Inspectors are designated "after consultation with the State or States concerned"; and, if the state so desires, the inspector "shall be accompanied by representatives of the authorities of the State concerned."[12] The apparently sweeping authority to have "access at all times" itself was limited in the statute: It was tied to data, places, and persons related to safeguarded material, equipment, or facilities and did not constitute a right for inspectors to roam freely in the territory of the safeguarded state. The most likely situation in which the rights elaborated in the statute would apply unvarnished would be those involving an agency project wherein the agency was itself the supplier and could insist on the safeguard terms and conditions for making assistance available in the first instance. And indeed, at the time of the drafting of the statute, expectations were high that much of the material and equipment under safeguards would in fact be supplied by the agency and that it would be appropriate under such circumstances for it to have very substantial oversight authority.

One of the potentially most important features of the safeguards rights included in Article XII is one that anticipates a need to establish effective means of control over plutonium. Sub-paragraph 5 accords the agency the right "to approve the means to be used for chemical processing of irradiated materials," although only to ensure the processing "will not lend itself to diversion" and not to determine the legitimacy of the activity itself. More importantly, the same subparagraph grants the agency the right "to require deposit with the Agency of any excess of any special fissionable materials recovered or produced" over what is needed for use for research or in reactors.

This provision, which also was included in anticipation of a substantial agency supplier role, has never been applied. But it has been central to the controversy in recent years about the merit and feasibility of international plutonium storage arrangements in conjunction with development of spent nuclear fuel reprocessing.[13]

These cautions and caveats notwithstanding, it is clear that the statute provides a broad and flexible basis upon which to develop and implement a safeguards system. Although the point will come up again later, it must be noted that the agency must be able to exercise whatever rights are necessary to ensure effective international safeguards. If those rights are deemed unduly curtailed, the agency could conclude that it is unable to apply safeguards in a manner sufficient to enable it to meet its responsibilities. In that event, the international community could conclude that international safeguards cannot be relied upon as a means of international verification. As discussed earlier, that conclusion would have heavy and far-reaching consequences.

The scope of the rights described underscores once more the unique qualities of international safeguards, and clarifies why negotiations over the safeguards principles to be incorporated in the statute and other instruments were "long, arduous, and often contentious."[14] On the other hand, it is just short of startling that at the statute conference that created the IAEA, among the eighty-one participants, no negative votes were cast on safeguards, and the highest number of abstentions totalled three, even though almost half of the time was devoted to debating safeguards and a substantial number of amendments were introduced and voted on.[15]

The Agency's Safeguards Document

IAEA safeguards are not self-executing and depend for implementation on an agreement between the agency and the nation concerned. The IAEA's safeguards system is embodied in two documents designated as INFCIRC/66/REV.2 (hereafter referred to as noted or as INFCIRC/66 or the agency's safeguards document) and INFCIRC/153 (hereafter referred to as noted or as the NPT safeguards document). The first document applies to safeguards arrangements negotiated pursuant to provisions of the statute of the IAEA and almost always involves only selected nuclear activities of the state involved; the second is concerned with safeguards applied to parties to the NPT and covers the entire peaceful nuclear programs of such states.

Technically, the agency's rights as defined in the statute could have been directly incorporated without change or elaboration in such agreements. Political conditions, however, were not conducive to such an approach, and it was necessary from the beginning to establish guidelines for the application of safeguards. This necessity had its own virtue. Since the agency, as an international organization, might find it difficult to be assertive in negotiating implementation of its rights with states that brandished their sovereignty, this approach offered greater promise of achieving a constructive outcome; a baseline could be defined for standardized and credible safeguards that even the most assertive states would find difficult to reject. The guidelines eventually took the form of INFCIRC/66, which served as instructions of the board of governors to the secretariat for negotiating safeguards agreements.

Agency safeguards have developed incrementally. INFCIRC/66 was itself a result of a review of an earlier safeguards document, INFCIRC/26, which had been adopted by the board of governors in January 1961 after extensive discussions that often were more political than technical in character.[16] This document was negotiated at a time when the political environment in the agency still put a strain on constructive cooperation in the field of safeguards. India, supported by the Soviet Union and some others (such as the United Arab Republic), had sought to impede progress in implementing safeguards, viewing them as fundamentally discriminating against developing nations.[17] INFCIRC/26 was designed to apply only to small reactors of less than 100 megawatts (thermal). This modest first step reflected not only the needs of the moment, but also recognized national sensitivities to safeguards activity and the importance of the IAEA's earning confidence before moving to the more intensive safeguards measures that would inevitably be required as larger and more complex nuclear facilities became subject to safeguards. Prudence dictated moving in small steps.

This first safeguards document also had a number of limitations, the most significant being that it did not include the crucial right of "pursuit"—the automatic application of safeguards to successively produced generations of nuclear material. When, three years later, in a changing political atmosphere, the board of governors decided to extend safeguards arrangements to large reactors (INFCIRC/26 ADD.1) the right of pursuit was incorporated into the safeguards system. Upon approving extension to large reactors, the board also mandated a general review of the system. This was the point of origin of INFCIRC/66.

By the time that INFCIRC/66 was under consideration (1965), opposition to development of a sound basis for implementing agency safeguards had all but vanished, primarily because of a shift in Soviet policy in support of safeguards that dated to 1963. Deprived of this important political support, India put its effort into seeking to preserve a few important objectives in what was clearly emerging as the agency's safeguards system. This trend toward constructive cooperation generally was reflected in the number and tenor of the meetings of the board (fewer and shorter meetings; the absence of dilatory behavior; and more serious technical discussions) and in the responses of the somewhat less disciplined general conference to the board's reports on its decisions on safeguards provisions. The general conference vote on the resolution taking note of INFCIRC/26 (1961) for example, was 43-19-2; the resolution supporting INFCIRC/26/ADD.1 (1964) was endorsed 57-4-6. But INFCIRC/66 (1965) and the two board decisions extending agency safeguards to reprocessing plants and to fuel fabrication and conversion facilities (1966 and 1968, respectively) won unanimous support. At a minimum, this evolution attests to the increasing political acceptance of basic international safeguards principles and procedures and to a readiness on the part of the agency members to deal with safeguards issues in constructive rather than polemical terms while nevertheless staunchly defending their national interests. As it turned out, the mid-sixties in many ways opened a decade of increased support for international safeguards and for the International Atomic Energy Agency.

The overwhelming proportion of IAEA safeguarding activities today derives from NPT obligations and comes within the framework of INFCIRC/153. There are, nevertheless, important reasons to understand INFCIRC/66. First, many features of the NPT safeguards document are drawn from its predecessor and from the experience accrued in its implementation. Therefore, INFCIRC/66 is relevant to understanding the NPT safeguards document. Second, while the preponderance of safeguards activities are carried out pursuant to NPT obligations, nine non-nuclear weapon states with significant nuclear activities are being safeguarded in whole or part on the basis of INFCIRC/66. This group of states includes India, Pakistan, Israel, and South Africa, all countries of proliferation concern to the international community that, along with Argentina, have at their disposal unsafeguarded as well as safeguarded nuclear facilities capable of producing weapons-usable material. (The other states that come under INFCIRC/66 safeguards are Brazil, Chile, Cuba, and Spain.) The importance of understanding the safeguards arrangements that apply in these threshold nuclear weapon states is self-evident. Finally, many of the states operating today under NPT safeguards arrangements earlier had negotiated safeguards agreements with the IAEA under INFCIRC/66. Those agreements were suspended at the time that new NPT agreements came into effect. In the event that an NPT state invokes Article X of the treaty, which permits withdrawal if a state "decides that extraordinary events, related to the subject matter of this Treaty have jeopardized the supreme interests of its country," the suspended safeguards arrangements of INFCIRC/66 would automatically be reinstated.[18] Hence, the non-NPT safeguards document remains germane, and could in the future assume even greater importance.

AN ANALYSIS OF INFCIRC/66

INFCIRC/66 consists of a basic document and two annexes that extend its provisions to reprocessing and fuel fabrication and conversion facilities. It has no provisions for enrichment or heavy water production facilities, although in 1967 the Soviet Union recommended that procedures be drawn up for uranium enrichment plants.[19] Plans for a meeting of experts from countries with enrichment experience were made but apparently superseded by the initiation of discussions leading to INFCIRC/153. The basic document consists of four parts:

- *General considerations*: covers the purpose and general principles of the document and of agency safeguards;
- *Circumstances requiring safeguards*: establishes the materials subject to safeguards as well as exemption, suspension, and termination provisions, including the manner of handling the transfer of safeguarded material out of the jurisdiction in which it is being safeguarded, and the knotty problem of substitution;
- *Safeguards procedures*: describes procedures in general and in respect to reactors and to materials outside principal nuclear facilities, including records, reports, and inspections; and
- *Definitions*: explains the key terms and concepts incorporated in the document.

General Considerations

Broadly defined, INFCIRC/66 is an instrument designed to establish a system of technical principles and procedures that permit the IAEA to comply with its statutory obligation to ensure that assistance or activities under its supervision or control are not used to further any military purpose. More specifically, its purpose is to verify compliance with safeguards agreements.[20] It emphasizes principle more than practice, providing a list of rules to govern the application of safeguards as well as the general approaches and procedures to be used. Unlike INFCIRC/153, which furnishes a model for a standard text of the structure and content of the safeguards agreements to be negotiated,[21] INFCIRC/66 establishes only a loose technical framework. While this can be viewed as a weakness, opening the possibility for determined states to seek to negotiate *de minimus* implementation arrangements, it also can be argued that the document serves as a statement of what is *minimally* acceptable; that it can be enhanced and improved but not derogated. Both positions have been argued, but on balance the trend in safeguards agreements negotiated under the umbrella of INFCIRC/66 has been toward arrangements that reflect changes in the political and technical environment. This suggests that the document's inherent limitations have not worked against the negotiation of constructive and useful safeguards agreements. If there is a major weakness in INFCIRC/66-based agreements, it is that they are not full scope. But that is because some states have been unwilling to accept full-scope safeguards, and is not the fault of the safeguards document itself.

It is worth emphasizing that agreements negotiated between the secretariat and a state must be approved by the board of governors, and that the secretariat is unlikely to present, and the board to accept, agreements that in their view do not provide for adequate safeguards. Indeed, in agreeing to some of the provisions of the safeguards document at the time it was before them for consideration, several board members made clear their view that the basic rights of the statute were in no way altered or diminished by virtue of the inclusion of certain formulations regarding the agency's safeguards rights and procedures.[22] This does not mean that INFCIRC/66 is free of difficulty. In contrast to INFCIRC/153, there was no formal provision for subsidiary arrangements to detail the specific measures that would be applied to particular materials and facilities. However, the secretariat quickly adopted the practice of using such arrangements to implement safeguards agreements.[23] Unlike the agreements, however, these subsidiary arrangements are not normally reviewed by the board and it is possible that the particular arrangements agreed to might involve some derogation of agency safeguards rights. That problem also could arise in the case of NPT safeguards agreements, although the standardization of subsidiary arrangements under that document significantly reduces the risk. (The lack of a specific provision for subsidiary arrangements is often cited as a deficiency of

INFCIRC/66 because these arrangements are seen as vehicles for the orderly arrangement of safeguards implementation.) These kinds of concerns have elicited recommendations in recent years for a generally more transparent international safeguards system. Other limitations of the agency safeguards document will be noted below.

Circumstances Requiring Safeguards

Safeguards, under INFCIRC/66/REV.2, may be applied to nuclear materials and to facilities in which such materials are used, processed, stored, or contained. The facilities included in this coverage are reactors for research and power production, spent fuel reprocessing plants, fuel fabrication and conversion facilities, and storage sites. INFCIRC/66 largely resolved an issue that was debated extensively in the early years of the agency: whether only nuclear materials or facilities as well were subject to safeguards. Principal suppliers of equipment favored applying safeguards to any assistance that could end up contributing to the production of fissionable material, but developing countries that were dependent on equipment supply and industrial states that eventually would be in a position to manufacture their own equipment favored restricting safeguards to the supply of nuclear material. In the end, it was agreed that it was primarily nuclear material that was subject to safeguards, but that facilities could be submitted to safeguards and that safeguards could be invoked even with respect to indigenous nuclear material if the latter were produced, processed, or used in a nuclear facility that had been substantially supplied to the recipient. In other words, while it was essentially nuclear material that was subject to safeguards, the supply of certain equipment, facilities, or non-nuclear material could invoke agency safeguards.[24]

The concept of "substantial supply" was central to the resolution of this question of the conditions under which safeguards would be brought into play. The first safeguards document had provided that agency safeguards would attach to "specialized equipment and non-nuclear material supplied by the Agency which in the opinion of the Board could *substantially assist* a principal nuclear facility."[25] This provision was offensive to a number of countries, and was one of the key targets of the review to which INFCIRC/26 had been submitted. Seeing that it would not be feasible to limit safeguards to transferred source and special fissionable material, India urged substituting the notion of "substantially supplied" for that of "substantially assisted." The rationale was that at least this would eliminate the triggering of agency safeguards merely as a consequence of providing know-how, financial assistance, design drawings and similar aid, and thus avoid intensifying what developing nations perceived as discriminatory arrangements in the first place.

Safeguards do not extend to uranium or thorium mines or mills, as these are considered ores and excluded from the definition of "source materials" in the IAEA statute. However, nuclear material "produced in or by the use of safeguarded nuclear material" is subject to safeguards.[26]

The obligation underlying INFCIRC/66 does not entail a commitment by the state to place all of its nuclear activities under safeguards, only those activities that have been expressly submitted to safeguards. It assumes the possibility that some nuclear material may be located in the safeguarded state that is not subject to agency safeguards, and contains provisions allowing for exemption and suspension of safeguards as well as for substitution of unsafeguarded for safeguarded nuclear materials under carefully specified conditions. For example, up to one kilogram of special fissionable material and specified amounts of natural and depleted uranium can be exempted from safeguards at the request of the state.[27] Similarly, plutonium produced in a reactor with a very low (less than 100 grams) annual plutonium production rate or in a reactor with less than three megawatts thermal maximum calculated power is exempted from safeguards.[28] In these cases, the agency's discretion is very limited, although a request that appeared to constitute an abuse of the exemption right could be challenged by the secretariat on the ground that its first obligation is to fulfill its statutory responsibilities.

In addition, a state may seek suspension of safeguards on up to one effective kilogram of special fissionable material for purposes of processing, reprocessing, testing, research, and development,[29] and, furthermore, may ask for suspension specifically "for the purpose of reprocessing . . . with the agreement of the Agency" provided that it substitutes for it an equivalent amount of other nuclear material, not previously under safeguards, the precise amount of such material being calculated according to agreed rules contained in the safeguards document. The objective of this arrangement is to ensure that temporary suspension for specified purposes does not result in an increase in unsafeguarded nuclear material from material subject to safeguards.[30] In contrast to the exemption provisions, the agency here has the right to approve or reject the request.

Safeguards Procedures

The provisions discussed above regarding the circumstances requiring safeguards are rather different from those of the NPT safeguards document, which covers all material in peaceful use in the safeguarded state. On the other hand, the actual safeguards procedures of INFCIRC/66 are quite similar not only to those of INFCIRC/153 but to those of its predecessor, INFCIRC/26. These include design review, records, reports, and inspection, all of which are specified in Article XII of the agency statute.

Recordkeeping and reporting are quite straightforward. Records provide a basis for reports submitted to the agency and are important in view of the fact that auditing and accounting procedures form the basis for international safeguards. While some criteria are common to all reports,[31] the precise arrangements are subject to negotiation between the state and the agency. In this respect, the procedures are less systematic than in the case of parallel procedures for NPT safeguards. Special reports are required in the case of unusual incidents entailing the actual or potential loss or destruction of a facility or material under safeguards and in the event of a transfer of safeguarded nuclear material.

According to Article XII.A.1 of the statute, the agency will examine the design of facilities and approve them to ensure that they will not serve any military purpose and that they will permit effective application of safeguards. The potentially far-reaching concept of approval never really took root and design review came to mean ascertaining that a facility was so designed as to permit effective safeguarding. The approval concept originally had been included in anticipation of an active program of agency projects wherein the IAEA, as supplier, would have to be sure that it was not supporting activities that would serve a military purpose. In fact, most safeguarding responsibility resulted from voluntary transfers where the key questions were whether safeguards could be effectively applied to the facility and whether the agency was satisfied that it could carry out its responsibilities. The agency's design review activity was diminished even further in the NPT safeguards document, which only provides that the IAEA be apprised of design information to enable it to determine the appropriate safeguards measures to be applied.[32]

Inspection is the most essential of the safeguards procedures in the sense that it is the means by which the credibility of the other elements can be established. It is also the most contentious because it is the most intrusive from the point of view of sovereignty. Even among supporters of international safeguards, many are inclined toward minimalism when it comes to formulating specific inspection arrangements. It is also true that the agency routinely schedules less inspection effort than facility attachments permit, and only carries out a portion of those planned. But this is in no small measure a function of resource limitations, not an inherent desire to minimize physical inspection.

INFCIRC/66 specifies inspection procedures in only a general way. A distinction is made between special and routine inspections, but the latter are the most important. The governing concept is that of inspection frequency (inspection intensity in the case of NPT safeguards) involving an inspection schedule based on annual usage or maximum potential production of designated nuclear materials. For nuclear reactors in which the annual throughput exceeds sixty effective kilograms, the IAEA has a right of access "at all times."[33]

In the case of reprocessing plants,[34] if the annual throughput does not exceed five effective kilograms, the frequency of routine inspection is fixed at two per year, but where it does exceed that number, the plant and the safeguarded material in it may be inspected at all times. For those plants having an annual throughput of more than sixty effective kilograms, "the right of access at all times would normally be implemented by means of continuous inspection."[35] Although continuous inspection is not defined in the safeguards document, discussions at the time of drafting indicate that this might entail resident inspection arrangements.[36] Similar provisions exist for fabrication and conversion facilities and for any other principal nuclear facility that the board of governors may designate, such as enrichment plants.

The maximum frequency of routine inspection formula states what *could* be done, and not necessarily what is actually done. Under INFCIRC/66, actual inspection frequencies take into account broader fuel cycle and programmatic considerations (such as the nature of the reactor and of the material used or produced in it and whether or not the state has reprocessing capability). Additionally, inspection rights are carefully circumscribed in the safeguards document in light of their political sensitivity and fears of economic losses resulting from intrusion into plant operation. For example, the number, duration, and intensity of inspections actually carried out "shall be kept to the minimum consistent with the effective implementation of safeguards."[37]

Sensitivity to Sovereignty

An important feature of international safeguards is that they developed and evolved with particular sensitivity to state sovereignty. The statute itself, as we have noted, limits the agency's right to exercise agency safeguards "to the extent relevant to the project or arrangement" in question. And those arrangements in turn are to be based on a safeguards agreement negotiated between the agency and the state concerned. The safeguards documents themselves contain further provisions designed to afford states assurance against the risk of arbitrary or capricious agency conduct in the implementation and administration of safeguards. Paragraph 4 of INFCIRC/66, for example, reaffirms that the safeguards provisions elaborated in that document come into force only with a safeguards agreement and then only to the extent relevant to the arrangement. Paragraph 15 of the document reiterates the point.

Numerous other provisions in the safeguards document emphasize obligations of the IAEA to states with whom it negotiates safeguards agreements. The agency is admonished: to implement safeguards in a manner designed to avoid hampering the economic or technological development of the state under safeguards;[38] to implement safeguards in a manner designed to be consistent with prudent management practices;[39] to take every precaution to protect commercial and industrial secrets;[40] and to ensure that no "commercial or industrial secret or any other confidential information" acquired by reason of safeguards implementation be disclosed except under designated circumstances and to designated individuals.[41] Some participants in the group responsible for drafting INFCIRC/66 would have preferred to go even further and to hold the agency absolutely responsible for protecting commercial and industrial secrets of which it acquired knowledge in the course of administering safeguards. Such a clause was not adopted, however. Outside knowledge of the "inputs" is limited because of the confidentiality of the subsidiary arrangements that implement the safeguards agreement; and outside knowledge of the "outputs" is limited because of the confidentiality surrounding inspection procedures. The net effect of these provisions is to not only provide confidentiality for the state being safeguarded, but to make it difficult for others to evaluate safeguards effectiveness.

Because of their even more manifest intrusion on sovereignty, inspection provisions are subject to additional constraints, including: the charge noted above to keep inspection frequencies and intensity to the minimum consistent with agency responsibilities; the specification of maximum inspection frequencies; the preclusion of inspectors from directing (in contrast with requesting) facility operators to carry out any particular operations; and the conditioning of certain activities (for example, conducting initial inspections to verify design review) on whether they are provided for in the safeguards agreement, thus making the inspection right somewhat dependent on the bargaining process. These constraints appear in even greater detail in the NPT safeguards document. When dealing with such questions as what opportunities inspection creates and what should be expected of safeguards, these constraints, and even more importantly what they reflect—state sovereignty—should be borne in mind.

Although INFCIRC/66 provides that the principles and procedures set forth are subject to "periodic review in the light of the further experience gained by the Agency as well as of technological developments,"[42] the safeguards document has never been reviewed. The reasons for this are largely political: the risk that opening up the document to review could lead to emergence of a weaker rather than a stronger instrument. INFCIRC/66 was negotiated in what retrospectively can be regarded as the halcyon days of the agency. Optimal political conditions existed for establishing an international safeguards system: convergent Soviet-American and Western interests, coupled with a generally euphoric view of the potentially substantial global benefits of the emerging age of commercial nuclear power, and the absence of any significant opposition on the board of governors, worked in favor of a robust agency and effective international safeguards. By the time experience in implementing the document had been gained, the NPT had come into force and divisions had appeared among the agency's membership according to whether or not the member was party to the NPT and had accepted nonproliferation undertakings and full-scope safeguards. Agency priorities, in the view of some states, became warped, and political confrontations over priorities, values, and resource allocation sharpened. These conditions were not conducive to achieving consensus on revisions that, if they were to strengthen the document, would have made it appear more rather than less like the more comprehensive safeguards document that had been fashioned to implement the agency's responsibilities under the NPT.

Improvements in Safeguards Procedures

INFCIRC/66 has not remained static in practice. Changes have occurred, in a few instances as a result of decisions of the board of governors, but more often because new provisions have been incorporated into the safeguards agreements pursuant to secretariat efforts, or to requirements insisted upon by suppliers as a condition of nuclear supply. This has helped to establish new precedents for future agreements. Nevertheless, it lacks the authoritativeness that would exist if the safeguards document itself were to assert these provisions as requirements. The most important of these changes are discussed here.

The director general in 1973, at the urging of a number of board members, proposed new rules regarding duration and termination of safeguards agreements negotiated under INFCIRC/66. This action was stimulated by the more comprehensive and enduring provisions of the NPT, and by the secretariat's having negotiated an agreement on Argentina's Atucha reactor, limited to five years with no mandatory provision for safeguards beyond that time. INFCIRC/66 made no explicit provision for dealing with nuclear materials after the expiration of a safeguards agreement. The assumption was that the project or arrangement upon which the agreement was based would contain appropriate provisions that would be incorporated into the safeguards agreement. As for the continuation of safeguards on special fissionable material produced as a result of use of safeguarded nuclear material, paragraph 16 of the safeguards document only spoke of the "desirability" of providing for the continuation of safeguards. Of particular concern were subsequent generations of special fissionable material that might be produced after a safeguards agreement had come to an end. The decision of the board to approve the director general's recommendation for dealing with these problems[43] significantly improved the situation. Since 1974, the duration of safeguards agreements has been tied to actual use in the recipient state of materials or items supplied. Safeguards continue on subsequent generations of produced nuclear material that is derived from originally safeguarded material or facilities until terminated in accordance with the termination provisions of the safeguards document. These rules apply not only with respect to the application of safeguards in connection with supplied nuclear material, but also to equipment, facilities, or non-nuclear material. In practice, this entails the perpetual application of safeguards. Ironically, this places non-NPT states under more stringent requirements than NPT states insofar as the latter are bound only for the duration of the treaty (initially, twenty-five years) and also have the right to withdraw under specified conditions.

Again at the level of the board, following an initiative by the secretariat, the agency in 1974 took the important step of redefining the undertaking of the safeguarded state. Until that time, states agreed not to use nuclear assistance to further any military purpose. Beginning in 1972, the United States and the United Kingdom declared that in their view "any military purpose" embraced any nuclear explosive devices and that their nuclear supply was predicated on the recipients undertaking not to use supplied items for such a purpose. In 1974, the director general proposed, and the board of governors accepted, an interpretation of "any military purposes" as including any nuclear explosive device.[44] Since 1975, board of governors approval of safeguards agreements negotiated on the basis of INFCIRC/66 has been contingent upon inclusion of an undertaking by the recipient that none of the supplied items covered by the agreement shall be used "for the manufacture of any nuclear weapon or to further any other military purpose[45] or for the manufacture of any other nuclear explosive device." In adopting this approach, the agency took a step in the direction of assimilating the non-NPT and NPT safeguards documents and in closing an avenue for acquisition of nuclear material outside of safeguards. In doing so, it strengthened the nonproliferation utility of international safeguards.

De Facto Changes in Procedure

A number of provisions found in safeguards agreements negotiated under INFCIRC/66 have no particular documentary antecedent, but have become accepted and approved by the board of governors and thus factored into the broader safeguards system. Since the mid-1970s, commencing with an agreement intended to cover supply to South Korea by France of a reprocessing plant (which was subsequently cancelled), safeguards agreements have included an undertaking by the recipient to accept safeguards on transferred technological information as well as on any nuclear facilities and equipment constructed or operated, or nuclear material produced, on the basis of such information. In cases involving sensitive technologies, there is, for a certain period of time, a conclusive presumption that any facility of the same type constructed contains transferred technology and is therefore subject to agency safeguards. It would be for the state constructing such a facility to demonstrate that it is not using technology of the type originally transferred.

Certain non-nuclear materials (such as heavy water and graphite), although not specified in the safeguards document, have come to be treated in much the same way as nuclear material, and are now both subject to safeguards and capable of triggering safeguards on nuclear materials and facilities with which they are associated. Similarly, heavy water production plants have been covered in safeguards agreements under procedures analogous to those designated to apply to nuclear material.

Unlike the NPT safeguards document, which specifically references containment and surveillance as "important complementary measures" to material accountancy, these safeguards methods are not explicitly mentioned in INFCIRC/66. Nevertheless, starting in the early to mid-1970s, they were introduced and today are routinely included in safeguards agreements.

The well-publicized case of Pakistan's Kanupp reactor demonstrates that neither inclusion of new safeguards measures not specifically provided for, nor the introduction of improved safeguards techniques can be taken for granted.[46] Pakistan only reluctantly accepted changes in safeguards procedures and demurred to changing the subsidiary arrangements to accommodate some improvements that the agency had identified as necessary to achieving effective safeguards. States that are not strongly supportive of safeguards are inclined to argue that amendments or modifications to safeguards agreements should be limited to changes made in the safeguards document itself or in the accompanying Inspectors Document (INFCIRC/39) pursuant to formal decisions by the board of governors. This approach seeks to elevate any adjustment or alteration in safeguards methods or procedures, even if based on agreed principles, to the political level where bargaining for other objectives can be sought as a concession for accepting technical adjustments, and to capitalize on the reluctance of many to subject the basic safeguards document to challenges that could weaken the international safeguards system. Predictably, strong supporters of safeguards seek to avoid desultory political debates and to define upgrading of safeguards agreements as technical matters to be resolved between the secretariat and the safeguarded state.[47]

SUMMARY ASSESSMENT

These few examples—related to the basic undertaking of the safeguards agreement, the duration and termination of safeguards, the treatment of technological information and of non-nuclear materials (such as heavy water and graphite), the methods for implementing safeguards, and the introduction of subsidiary arrangements as the administrative vehicle for carrying out safeguards responsibilities—

make it clear that INFCIRC/66 is neither static nor inflexible, and that in many respects it has been able to accommodate changing conditions in the nuclear environment. This is an important conclusion in view of the fact that this safeguards document will remain relevant for any state that chooses not to become a party to the NPT or the Treaty of Tlatelolco.

On the other hand, INFCIRC/66 has certain inherent weaknesses and is, by its nature, subject to some important limitations. Notwithstanding its apparent flexibility, INFCIRC/66 remains a framework, a set of principles, and not a model for an agreement. In this respect, it is different from the NPT safeguards document. Modifications and new provisions are introduced into safeguards agreements, and precedents presumably established thereby. But as additive elements to the safeguards document, rather than as formal amendments to it, they do not always carry the same authority as, for example, the incorporated elements of the NPT safeguards document, INFCIRC/153. This may be less true of the board's acceptance of the director general's legal interpretation of the meaning of "any military purpose" in the basic undertaking of a safeguarded state to include any nuclear explosive devices, but it is still the case that even that adjustment elicited the reservation of three governors and was not a formal board decision. Other adjustments introduced as technical measures by the secretariat are even more vulnerable.

The limitations to which INFCIRC/66 is subject derive not from the document itself but from external considerations. The fundamental problem is that the obligations that it covers constitute only partial nonproliferation measures. States that accept safeguards under this document have accepted only limited undertakings, agreeing only not to use certain specified items for military purposes (defined, as we have seen, to include any kind of nuclear explosive devices). They have *not* agreed to place all of their nuclear activities under international safeguards. They have *not* accepted the nonproliferation commitment associated with the NPT or with the Treaty of Tlatelolco, or any other comparable instrument, as have parties to those treaties. They are not pledged not to acquire nuclear weapons or explosives, and they remain legally and politically free to develop nuclear weapons or nuclear explosive devices from material, technology, and equipment that is not subject to international safeguards. What INFCIRC/66 was designed to do, it does well. The difficulties lie beyond the scope of its competence or authority. As long as unsafeguarded nuclear material and facilities exist, a window of vulnerability to proliferation remains open. Of course, for reasons addressed in chapter 1, even total acceptance of full-scope safeguards could not absolutely guarantee against further proliferation, but its adoption is one of the crucial steps that must be taken if the nonproliferation battle is eventually to be won.

NOTES

1. IAEA Statute, Article II.

2. Hans Grumm, "IAEA Safeguards—Status and Prospects," in *Nuclear Safeguards: A Reader*, report prepared by the Congressional Research Service, Library of Congress for the Subcommittee on Energy Research and Production, Transmitted to the Committee on Science and Technology, U.S. House of Representatives, 98th Congress, 1st Session (Washington, D.C., U.S. Government Printing Office, 1983) pp. 628–650; quote on p. 643.

3. "The Baruch Plan: Statement by the United States Representative [Baruch] to the United Nations Atomic Energy Commission, June 14, 1946," *Documents on Disarmament 1946–1959*, Pub. No. 7008, 2 vols. (Washington, D.C., U.S. Government Printing Office, 1960) vol. 1. p. 10.

4. First Report of the United Nations Atomic Energy Commission to the Security Council, December 31, 1946, in *Nuclear Safeguards: A Reader* (see note 1 above) pp. 62–89; quote on p. 86.

5. For the Atoms-for-Peace proposal see, "United States 'Atoms-for-Peace' Proposal: Address by President Eisenhower to the General Assembly, December 8, 1953," in *Documents on Disarmament* (see note 3 above) vol. 1 pp. 399–400; for the Atomic Energy Act of 1954 see Public Law 83-703, 68 Stat. 919, approved August 30, 1954, as amended.

6. For a general introduction and overview of safeguards as developed and applied by the IAEA, see IAEA, *IAEA Safeguards: An Introduction*, INF/SG/INF/3 (Vienna, IAEA, 1981).

7. On this point, see Rudolph Rometsch, "International Safeguards Objectives, Status and Unresolved Issues: The IAEA View," in *Nuclear Safeguards: A Reader* (see note 2 above) pp. 449–457.

8. Efforts to make safeguards acceptance a condition of membership in the IAEA were raised and rejected during negotiation of the statute. Much of what follows is based on a study (carried out by the author for International Energy Associates Ltd., Washington, D.C., under a contract with the U.S. Arms Control and Disarmament Agency [U.S. ACDA]), *Review of Negotiating History of International Atomic Energy Agency (IAEA) Document INFCIRC/66/REV.2, "The Agency's Safeguards System"* (U.S. ACDA, July 30, 1984).

9. IAEA Statute, Article XII.A.6.

10. Ibid.

11. Ibid., Article XII.A.1.

12. Ibid., Article XII.A.6.

13. On the effort to develop an international plutonium storage arrangement, see Charles N. Van Doren, *Toward an Effective International Plutonium Storage System*, A report prepared under a research contract for the Congressional Research Service, Report No. 81-255 S (CRS, November 1981). The question of statutory intent and coverage has been raised by James de Montmollin of Sandia National Laboratory in correspondence with the author and others at professional meetings on international safeguards. De Montmollin's main argument is that one should distinguish between what may be legally authorized in the event of a voluntary agreement by states to participate in an arrangement such as international plutonium storage, and the notion of discretionary authority of the agency to impose requirements on states.

14. Myron B. Kratzer, "Safeguards Against Nuclear Proliferation," in *Nuclear Safeguards: A Reader* (see note 2 above) pp. 253–266; quote on p. 254.

15. See Alan McKnight, *Atomic Safeguards: A Study in International Verification* (New York, UNITAR, 1971).

16. Ibid., chapter 3.

17. Bertrand Goldschmidt, *The Atomic Complex: A Worldwide Political History of Nuclear Energy* (La Grange Park, Ill., American Nuclear Society, 1982).

18. See INFCIRC/153, "The Structure and Content of Agreements Between the Agency and States Required in Connection with the Treaty on the Non-Proliferation of Nuclear Weapons," paragraph 24.

19. Alan McKnight, *Atomic Safeguards: A Study in International Verification* (see note 15 above) p. 58.

20. INFCIRC/66/REV.2, "The Agency's Safeguards System (1965, As Provisionally Extended in 1966 and 1968)," paragraph 46. Hereafter referred to as above or as INFCIRC/66 or as the agency's safeguards system.

21. Some people contend that INFCIRC/153 is not a model agreement because it does not provide a standard format but only lists provisions that normally should be included in a safeguards agreement. However, in 1974, at the request of the board of governors, the secretariat translated INFCIRC/153 into a standard text for agreements to be negotiated pursuant to the NPT. This was completed in August 1974 and published as GOV/INF/276, "The Standard Text of Safeguards Agreements in Connection with the Treaty on the Non-Proliferation of Nuclear Weapons" (Vienna, IAEA, August 22, 1974). The secretariat and members of the agency regard 153 as essentially a model agreement.

22. Based on discussions with participants in the negotiations.

23. See Paul Szasz, *The Law and Practices of the International Atomic Energy Agency* (Vienna, IAEA, 1970) chapter 21. Of course it is difficult to imagine safeguards agreements that did not include detailed provisions, whatever they might be called.

24. INFCIRC/66/REV.2 (see note 20 above) paragraph 19.

25. INFCIRC/26, "The Agency's Safeguards" (Vienna, IAEA, 1961) paragraph 37.

26. INFCIRC/66/REV.2 (see note 20 above) paragraph 19e.

27. Ibid., paragraph 21; this also appears in INFCIRC/153.

28. Ibid., paragraph 22; this also appears in INFCIRC/153. In addition, both safeguards documents provide for high enriched and low enriched uranium exemptions.

29. Ibid., paragraph 24.

30. Ibid., paragraph 25. "Effective kilogram" is a concept introduced to establish equivalences of different levels of enriched uranium. There is no differentiation for plutonium.

31. INFCIRC/66/REV.2 (see note 20 above) paragraph 39.

32. INFCIRC/153 (see note 18 above) paragraphs 42, 46.

33. INFCIRC/66/REV.2 (see note 20 above) paragraph 57. This principle of inspection frequency derives from the statute and was first implemented at the time that INFCIRC/26 was extended to cover reactors larger than 100 MW. See INFCIRC/26/Add.1 (April 1964).

34. Annex I (so-called REV.1) of INFCIRC/66 (see note 20 above).

35. The statement about continuous inspection appears only in a footnote to INFCIRC/66 (see note 20 above), reflecting the political sensitivity of the concept of continuous inspection. Agreement could not be reached to incorporate the concept in the body of the paragraph providing for specific inspection arrangements. Hence, the relevant paragraphs in both annexes (REV.1 for reprocessing plants and REV.2 for conversion and fabrication plants) say only that in cases involving more than five effective kilograms of nuclear material, the plant and nuclear material in it "may be inspected at all times." The footnote explains that where sixty or more effective kilograms are involved, the right of access at all times would "normally be implemented by means of continuous inspection."

36. Ibid.

37. Ibid., paragraph 47.

38. Ibid., paragraph 9.

39. Ibid., paragraph 10.

40. Ibid., paragraph 13.

41. Ibid., paragraph 14.

42. Ibid., paragraph 8.

43. IAEA document GOV/1621, "The Formulation of Certain Provisions in Agreements under the Agency's Safeguards System (1965, As Provisionally Extended in 1966 and 1968)" Memorandum by the Director General (Vienna, IAEA, August 20, 1973).

44. See IAEA, *International Safeguards and Non-Proliferation of Nuclear Weapons*, page 15, where it is noted that three members of the board of governors expressed reservations concerning this interpretation.

45. It is this phrase that excludes use of nuclear material for naval propulsion units in submarines.

46. On this issue, see David Fischer and Paul Szasz, *Safeguarding the Atom: A Critical Appraisal* (London and Philadelphia, Taylor and Francis, 1985) chapter 3.

47. See Benjamin N. Schiff, *International Nuclear Technology Transfer: Dilemmas of Dissemination and Control* (Totowa, N.J., Rowman & Allanheid; London, Croom Helm, 1983).

Chapter 5

NPT SAFEGUARDS

In transmitting the Treaty on the Non-Proliferation of Nuclear Weapons (NPT) to the U.S. Senate for its advice and consent in 1968, President Lyndon Johnson referred to it as "the most important international agreement limiting nuclear arms since the nuclear age began."[1] That treaty consigns responsibility to the International Atomic Energy Agency (IAEA) for verifying that non-nuclear weapon state parties fulfill their obligation not to use their peaceful nuclear activities to develop nuclear explosive devices of any kind. It is therefore not surprising that the NPT has had a deep and lasting effect on the IAEA and on international safeguards. In this chapter, we will examine the nature of the NPT-related safeguards and the ways in which agency safeguards were adapted to meet the responsibilities assumed under the NPT.

NATURE OF THE CHANGE

There is a basic difference between the undertaking assumed by non-nuclear weapon state adherents to the NPT and the undertakings discussed in the previous chapter. When non-nuclear weapon states decide to become parties to the NPT they formally renounce nuclear weapons. In ratifying the NPT, these states undertake not to receive, manufacture, or otherwise acquire nuclear weapons or other nuclear explosive devices. In addition to accepting the obligation not to manufacture nuclear weapons or nuclear explosive devices these states also commit themselves to accept safeguards "as set forth in an agreement to be negotiated and concluded with the International Atomic

Energy Agency . . . [on] all source or special fissionable material in all peaceful nuclear activities within the territory of such State, under its jurisdiction, or carried out under its control anywhere."[2] As discussed in chapter 1, this sweeping commitment is subject to an exception (never yet invoked) in the case of nuclear materials declared as being used for nonweapons, nonexplosive military purposes such as in naval submarine reactors. This exception notwithstanding, the nonproliferation commitment of the NPT and the acceptance of comprehensive safeguards on all peaceful nuclear activities significantly augment the political and security relevance of the IAEA and of international safeguards.

In contrast, non-nuclear weapon states that are not party to the NPT or Treaty of Tlatelolco, have only pledged not to use certain designated items for any military purpose, and to accept international safeguards on those items as well as on special fissionable material produced as a result of their use. However, they remain free to pursue other nuclear activities that are not subject to any restraint or obligation, and to which safeguards need not be applied. It is thus both possible and legally permissible for them to maintain unobligated and uncontrolled nuclear activities parallel to those under obligation without in any way violating their undertakings and commitments. As we already have noted, a number of important states today operate both safeguarded and unsafeguarded nuclear facilities and are thus in a position to acquire weapons-usable nuclear material that is not subject to any constraint or safeguards.

EXPANDING THE SCOPE OF SAFEGUARDS

The NPT has involved not only increased political relevance, but also a very significant expansion of the IAEA's safeguarding activities. The 1970 safeguards budget amounted to slightly over $1 million or 10 percent of the assessed budget of the agency. By 1980, it had grown to $19 million, or 25 percent, and to $32.5 million, or 33.6 percent, by 1984.[3] Even after including voluntary technical cooperation funds and taking a percentage of the total budget, the same upward trends apply, although at a slightly reduced level (about 8.5 percent in 1970; 21.5 percent in 1980; and 27.1 percent in 1984). Perhaps the most graphic expression of the impact of increased safeguards responsibilities (due primarily but not entirely to NPT) on the agency can be seen in table 5-1, which indicates the dollar ratio in terms of program expenditure between safeguards and all

TABLE 5-1. COMPARISON OF EXPENDITURES FOR SAFEGUARDS AND ALL OTHER PROGRAMS
(absolute numbers in million U.S. dollars)

Category of expenditures	Years				
	1958–62	1963–67	1968–72	1973–77	1978–82
Safeguards	1.3	1.9	7.1	27.2	109.9
All other programs	9.2	14.0	20.5	52.8	116.7
Ratio	1:7	1:7	1:3	1:2	1:1

Source: Derived from *Review of the Agency's Activities*, GC (XXVIII)/718 (Vienna, IAEA, July 1984).

other agency programs. As can be seen, the ratio diminished from 1:7 in the first two periods (1958–1962 and 1963–1967) to 1:3 in the transition period when NPT safeguards began to take effect. It declined to 1:2 and 1:1, respectively, in the periods 1973–1977 and 1978–1982.

Among other things, this change reflected the increase in nuclear power programs and the fact that growing peaceful nuclear activities in most non-nuclear weapon states were encompassed by the NPT. A change had occurred from safeguarding individual facilities in a relatively small number of states to safeguarding all nuclear activities in a larger number of states. Thus, whereas in 1970 only 10 power reactors and 4 bulk handling facilities were under agency safeguards (a number that increased to more than 150 when research reactors, storage sites, and research facilities were included), 126 power reactors and 48 bulk handling facilities (and a total of 772 nuclear facilities or locations outside facilities) were being safeguarded in 1980.[4] By 1984, a total of 875 installations of all kinds were under IAEA safeguards. In addition, ten facilities in nuclear weapon states were under agency safeguards.[5]

By 1984, the volume of nuclear material covered by these safeguards had grown to 129.5 tons of plutonium contained in spent nuclear fuel and 7.7 tons of separated plutonium (including some plutonium located in nuclear weapon states and submitted under voluntary safeguard arrangements), a substantial increase over the 83 tons of contained and 6 tons of separated plutonium (none in weapon states) in 1980, and only 770 kilograms of plutonium in all forms in 1970. The pattern is the same for highly enriched uranium (11.8 tons in 1984, 11 tons in 1980, and considerably less than a ton in 1970); low enriched uranium (22,784 tons in 1984, 13,872 tons in 1980, and 243 tons in 1970); and for source material (31,724 tons in 1984, 19,097 tons in 1980, and 1,146 tons in 1970). In addition, 1,362 tons of heavy water were under IAEA safeguards in 1984.[6]

To put this in perspective, by the early 1980s, 97 percent of the nuclear plants in all non-nuclear weapon states were under agency safeguards (86 percent under NPT arrangements and 11 percent under non-NPT arrangements). As for inspectors and number of inspections carried out by the IAEA over the course of this fifteen-year period, the situation is as follows: in 1970, a total professional safeguards staff of 54 (of whom only a modest number were accredited inspectors) conducted approximately 180 inspections; in 1980, a total professional safeguard staff of slightly more than 200 (of whom about half were accredited inspectors) conducted 1,100 inspections; and in 1984, 154 accredited inspectors (out of a staff of 230 professionals) conducted 1,820 inspections. Between 1968 and 1982, the IAEA actually carried out a total of 9,600 inspections, nearly two-thirds of which occurred in the period between 1978 and 1982; during this period, the major industrial NPT states, including EURATOM countries and Japan, had completed the vast majority of their safeguards arrangements with the agency.

IMPACT OF NPT SAFEGUARDS RESPONSIBILITY ON IAEA COHESION

Beyond the enhanced general political relevance of IAEA, and the substantial expansion of its responsibilities deriving from linkage to arms control through the NPT, still another effect on the IAEA can be identified: the impact of NPT responsibilities on the cohesion of the organization. Although this is a subject that will be treated in greater detail in chapter 7, a few observations are in order here as background to the discussion of the agency's NPT-related safeguards document.

Since not all members of the IAEA have chosen to join the NPT, two groups of states have emerged in the agency, each subject to a different set of safeguards arrangements reflecting the different undertakings of the respective groups and the consequential extent to which their nuclear activities are subject to safeguards. Over time, pressures have developed to achieve greater convergence between the two sets of arrangements; the balance is weighted in favor of the more comprehensive NPT safeguards. We have already discussed several examples: the extension of the basic undertaking in INFCIRC/66/REV.2 "not to use certain items in such a way as to further any military purpose" to include the manufacture of any nuclear explosive device; the inclusion of containment and surveillance in INFCIRC/66 safeguards agreements as standard safeguards methods; and the requirement of GOV/1621 that safeguards agreements remain in force until safeguards have been terminated on all items subject to safeguards—a requirement that reflects the spirit though not the letter of the NPT.

Actions such as these have evoked strong criticism by non-NPT agency members. Their position is that the introduction into safeguards agreements, as standard features, of requirements arising out of the NPT distorts both the letter and the spirit of the statute. This has contributed to a weakening of the consensual base upon which the agency had operated throughout the greater part of its existence, and has stimulated the emergence of more confrontational politics than is healthy for an organization with as novel, sensitive, and potentially far-reaching implications for international stability and security.

Another spinoff has been the emergence of the view among a substantial part of the membership that an increasingly inappropriate balance was developing between technical assistance and safeguards activities as reflected in the progressively larger proportion of total agency resources allocated to safeguards. This in turn has prompted a debate over whether the statute contains an inherent requirement of balance between its statutory responsibilities to promote nuclear energy and to ensure against its abuse.

Finally, the attention given to the agency because of the assignment to it of important, contentious responsibilities with high political content (that is, safeguards) has, in the view of some, made the agency an increasingly attractive forum in which to conduct political campaigns regarding issues only marginally, or sometimes not at all, associated with its mandate. In their view, the importance attached by some member states to effective international safeguards creates the potential for safeguarding activities to be held hostage to securing satisfaction on other issues ranging from politically isolating and attacking certain states (like Israel and South Africa), to challenging western technology exchange policies, to achieving technical cooperation and economic development goals of interest to them. While some of these objectives are widely accepted as clearly legitimate, others are not. The need to reckon effectively with these challenges is evident. For the present, however, the point is only to underscore the diverse effects of the IAEA's responsibility to implement NPT safeguards.

NPT SAFEGUARDS PROVISIONS

Article III.1 of the NPT requires non-nuclear weapon state parties to "accept safeguards, as set forth in an agreement to be negotiated with the International Atomic Energy Agency in accordance with the Statute . . . and the Agency's safeguards system. . . ." The more comprehensive nature of the safeguards required under the NPT—covering all of a state's peaceful nuclear activities—stimulated demands by a number of industrial states with advanced nuclear capabilities that the agency's safeguards system be reviewed in light of that broader mandate. The Federal Republic of Germany and Japan, for example, were particularly concerned that adherence to the NPT not impair their peaceful nuclear development programs or place them at competitive disadvantage in the anticipated world nuclear market, particularly in respect to such sensitive fuel cycle activities as enrichment, reprocessing, and development of breeder reactors. To deal with this concern, President Johnson offered to permit the IAEA to apply safeguards to all nuclear activities in the United States other than those with direct national security significance—an offer that took effect in 1980. For differing reasons, the United Kingdom, France, and the Soviet Union made similar offers that were implemented in 1978, 1981, and 1985, respectively, and the People's Republic of China is currently negotiating an implementation agreement for a similar plan. Indeed, it was rather clear that ratification of the NPT by Japan and the Federal Republic of Germany was contingent on the establishment of satisfactory safeguards arrangements. In their view, the system represented by INFCIRC/66/REV.2 lacked sufficient certainty and predictability to ensure against unnecessary intrusiveness, especially with respect to the issue of inspection intensity.[7]

The board of governors, itself not quite certain how to implement its new mandate, established a committee open to all agency members to advise it on the agency's responsibilities in relation to the NPT safeguards provisions and to determine the content of safeguards agreements with NPT parties.[8] The result of this effort, which involved 82 committee meetings over a 10-month period, was INFCIRC/153, "The Structure and Content of Agreements Between the Agency and States Required in Connection with the Treaty on the Non-Proliferation of Nuclear Weapons." This document and INFCIRC/66/REV.2 constitute today's agency safeguards system.

INFCIRC/66/REV.2 and INFCIRC/153 Compared

The two agency safeguards documents have several similarities. Both rely for implementation on the same methods and practices including records and reports and the use of design information to establish safeguards strategies. Both give primacy to on-site inspection as the key means of achieving independent verification. Both documents also emphasize that the agency will not hamper facility operations or impede peaceful nuclear development in implementing safeguards; it will employ prudent management practices; and it will not reveal proprietary or confidential information obtained by the agency in the course of conducting its safeguards activities.[9] INFCIRC/153 is even more vigorous in this respect: for example, it contains specific provisions regarding the designation of agency inspectors and protecting a state's right to refuse a particular designation. However, it also characterizes a state's repeated refusals to accept designated inspectors as an effort to impede the effective application of safeguards that is subject to action by the board of governors,[10] and it establishes more elaborate procedures than did INFCIRC/66/REV.2 for interpretation and application of safeguards agreements and resolution of any ensuing disputes.[11]

Some significant differences also exist. These are reflected in a number of new concepts introduced as elements of the NPT safeguards document with a view to improving the acceptability and effectiveness of safeguards in light of the agency's broadened mandate. These new concepts include: formal provision for subsidiary arrangements; emphasis on focusing safeguards on strategic points;[12] the use of instrumentation and other nonhuman inspection techniques where feasible;[13] explicit recognition of surveillance and containment as important complementary measures to material accountancy;[14] a requirement that safeguarded states establish national systems of accounting for and control of nuclear material;[15] and the regulation of the intensity of routine inspection through level of effort rather than through inspection frequency, with actual inspection intensity being determined in light of such factors as the degree of functional independence of the facility operator from the state control system and "the extent to which the State's nuclear activities are interrelated with those of other States."[16] In general, INFCIRC/153 sets tighter limits on what the IAEA can do, thereby providing protection for states against escalation of safeguards in the future; this had been a matter of considerable concern to a number of non-nuclear weapon states at the time the NPT safeguards were elaborated.

INFCIRC/153 consists of three parts, each of which is briefly described below:

Part I establishes the general principles governing the rights and obligations of the parties (state or states and the agency) to safeguards agreements under the NPT. This includes the basic undertaking, principles of implementation of safeguards, legal and financial responsibilities, measures that can be taken by the agency in an effort to ensure verification of nondiversion, and rules for interpretation and settlement of disputes.

Part II translates these principles into operational arrangements and spells out the provisions necessary to implement safeguards. It includes a statement of the technical objectives of safeguards, the materials to be covered, the measures to be applied (material accountancy, containment, surveillance), and the methods to be used (design review, records, reports, and inspections).

Part III consists of definitions of key terms and concepts (such as *material balance area, material unaccounted for*), and strategic points that appear in the principles and operational procedures sections of INFCIRC/153.

SIGNIFICANT FEATURES OF NPT SAFEGUARDS

Subsidiary Arrangements

In contrast with INFCIRC/66/REV.2, INFCIRC/153, as its title—"The Structure and Content of Agreements"—suggests, is more than a general framework of safeguards procedures and is much more a model for safeguards agreements. Even so, its provisions are still rather general, leaving specific details and implementation provisions in particular cases to be elaborated in the safeguards agreement and especially in its associated documents.

Three negotiated instruments are involved in the implementation of INFCIRC/153 safeguards: the Safeguards Agreement, Subsidiary Arrangements, and Facility Attachments (which include a Design Information Questionnaire). Other than the safeguards agreement itself, which establishes the basis for the application of agency safeguards, the most important of these instruments is that which deals with subsidiary arrangements.

The subsidiary arrangements specify the details of safeguards implementation according to the general principles in the safeguards agreement.[17] They define what information the agency will require to meet its safeguards obligations, establish a format for recordkeeping, and delineate the safeguards arrangements to be carried out at each facility. They include the facility attachments, which in turn contain definitions of material balance areas and the strategic and key measurement points at which safeguards will be applied in normal circumstances; the containment and surveillance measures to be applied; the format and timing of reports to be submitted to the IAEA; and the mode, timing, and extent of IAEA inspection activities at the facility. In short, they contain all of the crucial elements of a safeguarding exercise.

One of the key objectives of these arrangements is to facilitate standardization in the application of safeguards while preserving confidentiality of information provided to the agency, consistent with the general principles elaborated in Part I of INFCIRC/153. In practice, subsidiary arrangements are negotiated between the agency and the state to be safeguarded; they are not imposed by the agency on the state. Unlike the safeguards agreements, the subsidiaries are not submitted to the board of governors for discussion or approval, and their contents are not published or in any other way publicized.[18]

The agency must be satisfied as to the effectiveness of the safeguards that can be applied under the negotiated arrangements. If it were otherwise, the provisions of the safeguards document would have no rationale. This does not mean that changes and adjustments may not need to be made, for example to accommodate advances in safeguards instrumentation and technology, and in fact, paragraph 39 provides for the possibility of changing subsidiary arrangements by agreement between the state and the agency without need to amend the safeguards agreement itself. However, since many crucial and sensitive elements of actual safeguards arrangements are established in the subsidiary arrangements, including the definition of strategic and key measurement points and the designation of actual routine inspection efforts at different facilities, it is somewhat unclear whether modifications and adjustments can be readily achieved or whether the secretariat might have to consider invoking the authority of the board in an effort to bring about desired changes. This in turn raises the important question of whether, and under what circumstances, the board would feel moved as a collectivity to confront a member state. Even though it arose with respect to a non-NPT state operating under INFCIRC/66, the difficulties experienced in seeking to alter earlier understandings with Pakistan regarding the application of surveillance equipment at the Kanupp reactor suggest that agreement to change may be much more difficult to achieve than normally believed.[19]

In conclusion, subsidiary arrangements provide a basis for standardization and also a means by which the agency can exercise flexibility in the actual application of safeguards measures. Because they are negotiated instruments, they also carry the risk that states intent on doing so could pressure the agency to accept *de minimus* arrangements that might undercut the agency's implementation of its full safeguards rights. There is some evidence that this has occurred.

Strategic Points

INFCIRC/153 defines a strategic point as the location "where, under normal conditions, and when combined with the information from all 'strategic points' taken together, the information necessary and sufficient for the implementation of safeguards measures is obtained and verified. . . ."[20] It is also provided that in implementing safeguards the agency "shall make every effort to ensure . . . the application of the principle of safeguarding effectively the flow of nuclear material subject to safeguards . . . by use of instruments and other techniques at certain strategic points *to the extent that . . . technology permits*."[21]

The principle of safeguarding the flow of fissionable material at certain strategic points did not originate with INFCIRC/153. It had surfaced earlier at the initiative of the Federal Republic of Germany in 1967 when the board of governors was considering the extension of agency safeguards to conversion and fabrication facilities (INFCIRC/66/REV.2); and it appears in the preamble to the NPT itself. Interest in the strategic-points approach partly reflected concern that safeguards be cost-effective and efficient, and that the risk of exposure of proprietary information be kept to a minimum. It also bespoke a desire to make safeguards as nonintrusive and nonvisible as possible. The same concerns explain the parallel emphasis on introducing instrumentation in lieu of human inspection wherever feasible. Strategic points are justifiable as a basis for efficient, standardized safeguards approaches, and they are logically consistent with the concept of material balances, which is the primary method used in international safeguards. They offer a rational approach for dealing with the IAEA's augmented safeguards responsibility by emphasizing the concentration of safeguards effort where material is most accessible—presumably the places were diversion might most readily take place.[22]

The most important application of this principle arises in the context of inspection arrangements. INFCIRC/153 provides that access for purposes of routine inspections (that is, the normal and most common type of inspection) is limited to "strategic points specified in the Subsidiary Arrangements. . . ."[23] This contrasts sharply with the generalized statement in Statute XII.A.6 that agency inspectors have access as required "at all times to all places and data and to any person" in the safeguarded state and with the open-endedness of INFCIRC/66.

While strongly endorsing the strategic points concept, INFCIRC/153 nevertheless contains a number of constraints that protect the spirit of the statutory provision just mentioned. For instance, as noted above in the italicized language, the strategic points principle and the use of instruments and other techniques are conditioned by what is technologically feasible, consistent with the agency's need to be able to fulfill its safeguarding responsibilities. Moreover, as paragraph 116 asserts, the selection of strategic points must be sufficient to enable access by the agency to all of the information needed for effective safeguards. Furthermore, the safeguards document contains procedures for invoking special inspections entailing broader access rights if the agency finds that information obtained from routine inspections and made available by the state is inadequate for it to meet its verification obligations.[24]

Finally, there is the possibility of reassessing and revising strategic points (or other provisions, such as the designation of material balance areas or procedures for taking physical inventories). The normal basis for the selection of strategic points is by review of design information made available to the agency by the state. The NPT safeguards document provides in paragraph 47 that the selection of strategic points should be reexamined in light of changed operating conditions, safeguards technology developments, or experience in the application of verification procedures. However, as noted earlier, strategic points and other matters incorporated in the subsidiary arrangements are not imposed by the agency on the state or vice versa, but are in effect negotiated by the parties. This suggests that changes are in the final analysis the result of mutual agreement. Ipso facto, the same observations apply to questions related to strategic points and to the use of instrumentation.[25]

To summarize: strategic points offer greater specificity, certainty, and predictability, and thus serve the objective of improving safeguards acceptability. The agency is still left with a range of measures (special inspections, action by the board of governors) that could be invoked to provide up to full statutory access to safeguarded material. This, of course, depends on how prepared the agency is to take full advantage of its rights and to face the possibility of escalating confrontation with a delinquent or noncooperative state. In no case has the agency had to take matters to the point of confrontation (even the Pakistan case was resolved with strong urging but no formal action) but has managed to secure necessary access to be able to make a finding regarding diversion of nuclear material. This fact attests to the generally satisfactory nature of the approach.

State Systems of Accounting and Control

Another feature that distinguishes INFCIRC/153 from the earlier safeguards document is the requirement that states under NPT safeguards establish and maintain a system of accounting for and control of nuclear material.[26] The inclusion of such a requirement reflected practical and political considerations. Practically, it was not feasible for the IAEA to run a self-sufficient accountability system and to directly verify all flows and inventories. Although INFCIRC/66 does not require the establishment of state systems, it does call for agreement between the IAEA and the state on a system of records and reports. Virtually all states with significant nuclear activities have instituted control arrangements because of the value and hazardous nature of the materials in question, and the IAEA has made use of the records and reports generated under those systems.

The NPT safeguards document sets forth the basic elements that should be included in a state system of accounting and control (SSAC) in order to provide a basis for the application of IAEA safeguards. Among other controls, it calls for the establishment of a nuclear materials measurement system, procedures for taking physical inventories, a system of records and reports, and procedures for submitting reports to the IAEA.[27] States are not required to incorporate all or even any of these elements (although they normally do), but as other provisions of INFCIRC/153 make clear, the quality and technical effectiveness of the state system are relevant to the intensity of the agency's own safeguards effort. The closer the SSACs are in form and operation to the measures recommended, the less intensive the safeguards measures the agency would deem it necessary to impose. The agency verifies the findings of the state's control system, maintaining and implementing the right to make its own independent measurements and observations, and to supplement information provided by the state to assure itself that diversion has not taken place and that safeguarded materials are accounted for. Thus, whatever the technical effectiveness of SSACs, they do not qualify as substitutes for or alternatives to IAEA safeguards.[28] They are an integral part of the IAEA safeguards concept, but separate and distinct from the operational system directly controlled by the agency.

The political considerations related to the provision for national accounting and control systems are somewhat more complex, but also largely concerned with the issue of how to minimize the intensity of agency safeguarding activities. As noted, INFCIRC/153 states that in its verification, the agency "shall take due account of the technical effectiveness of the State's system."[29] The same admonition is repeated in paragraph 31, which also calls on the agency to avoid unnecessary duplication of the state's accounting and control activities. The provisions dealing with the critical question of determining the "actual number, intensity, duration and mode of routine inspections of any facility" assert that one of the criteria to be considered is the "effectiveness of the State's accounting and control system, including . . . the extent to which the measures specified in paragraph 32 above have been implemented by the state. . . ."[30]

It is noteworthy that the qualifying term "technical" does not appear before "effectiveness" in this paragraph. This suggests the possibility of a broader definition of the latter term and raises an historically important issue that continues to be relevant today: the status of the multinational regional EURATOM community. The question of how EURATOM as an existing and operating safeguards system would fit into the projected arrangement for international safeguards under the NPT arose at the time the treaty was being negotiated. EURATOM was unwilling to disband itself in favor of the IAEA, and some countries, in particular the Soviet Union, regarded EURATOM as anything but functionally equivalent to an international safeguards system or agency.[31]

A first step toward resolving this problem was taken by incorporating (Article III.4 of the NPT) a proviso that non-nuclear states could fulfill their obligation to conclude safeguards arrangements with the IAEA "individually or together with other states." This ultimately enabled the EURATOM non-nuclear weapon states (Federal Republic of Germany, Italy, Belgium, Netherlands, and Luxembourg) to authorize EURATOM to conclude a single agreement on their collective behalf. Of course, this still left open the question of how to implement those safeguards, and ratification of the NPT by the European Community's non-nuclear weapon states was predicated on achieving a satisfactory arrangement with IAEA that took due account of the EURATOM safeguards system.

It is an open secret that IAEA-EURATOM relations since 1970 have been strained. Although a safeguards agreement was concluded in mid-1972, it took five years to finalize the subsidiary arrangements, and even then some questions were left partially unresolved and lingered into the 1980s. The agreement reached involved a coordinated approach in which EURATOM acts as an agent of its member states for purposes of generating and submitting reports to the IAEA, and as an agent of the IAEA in the actual conduct of inspections, although under the surveillance of agency inspectors. Where sensitive facilities are involved, inspections are conducted jointly and the agency in any event reserves its statutory safeguards rights to conduct additional inspections if the circumstances require.[32]

It is interesting to speculate on just how the criterion of "effectiveness of the state's accounting and control system" should be interpreted, especially with its companion element, "including the extent to which the operators of facilities are functionally independent of the State's accounting and control system." Occasional comments from EURATOM suggest that this entails political as well as technical effectiveness, and that a regional arrangement such as EURATOM (wherein there is a clear separation between national level operators and supranational safeguards authorities) is so fundamentally distinctive that it is entitled to special treatment.

But this point of view is not universally shared. As an international organization the IAEA would be the first to acknowledge that it lacks capacity to make political judgments that patently discriminate among its members. Of equal relevance is the viewpoint of one of the members of the Japanese team that participated in the drafting of INFCIRC/153. Writing shortly after completion of the negotiations on the NPT safeguards document, Ryukichi Imai noted:

> It can be argued that multinational regional safeguards are credible while national safeguards are not. This involves a misunderstanding. What is being evaluated here is the mathematical formulation of the [national] system. . . . This proposition is quite different from a recommendation to accept the results of national or regional verification at their face value. . . . There is a very real concern that this misunderstanding may allow the five NNWS [non-nuclear weapons states] in the EURATOM organization to use their own safeguards system . . . and to accept only token verification by the Agency. This . . . would create inequality among NNWS. . . .[33]

A limited consensus seems to have emerged that technically effective accounting and control systems and functional independence of operators can have the effect of reducing the intensity of verification by the agency, but only insofar as it does not undermine the agency's fundamental right to independently verify nondiversion of nuclear material. Whether effectiveness is to be given a broader definition likely depends on whether the state taking a position has the agency's agreement that in light of its total situation (and not merely the technical efficacy of its accounting system) the state is entitled to special treatment.

In sum, the provision for SSACs assists the agency in carrying out its safeguards responsibilities within the realistic limits of its resources, and also buffers facility operators from the agency and gives the state a significant role in the safeguards effort. However, as the EURATOM case demonstrates, states might also seek to invoke SSAC effectiveness as a reason not only to reduce the intensity of the inspection effort, but also to curtail the scope of independent verification by the agency, thus creating tensions and problems of the type discussed in dealing with strategic points and subsidiary arrangements.

Regulating Routine Inspection Intensity

This brings us to a fourth closely related highlight of NPT safeguards—the criteria for regulating the intensity of inspection. Inspection, as we have noted, is the most sensitive issue associated with safeguards, for it is the procedure that most directly touches the nerve of sovereignty. The IAEA statute, it may be recalled, gives the agency broad rights of "access at all times to all places and data" required to determine compliance with relevant undertakings. INFCIRC/66 somewhat diluted that sweeping statement by providing that "the number, duration and intensity of inspections actually carried out shall be kept to the minimum consistent with the effective implementation of safeguards. . . ."[34] To further define inspection intensity, that document established a scale of maximum routine inspections based on effective kilograms of nuclear material in the safeguarded facility or its inventory,[35] and in the case of reactors specified several criteria for guiding *actual* routine inspection frequency.[36]

Similarly, the NPT safeguards document establishes the maximum routine inspection effort for designated categories of facilities or other locations of nuclear materials.[37] Based on the quantity of nuclear material involved, it specifies the number of man-days of inspection per year that is applicable to each designated category, leaving the agency to decide how to deploy that aggregated effort among the different facilities in each category in the safeguarded state.

However, all of this is conditioned by the same general constraint that appeared in the earlier document (INFCIRC/66); routine inspections are to be kept to the minimum consistent with the agency's responsibilities.[38] The *actual* routine inspection effort, which in practice is agreed (that is, negotiated on a facility-by-facility basis between the state and the agency and included in the subsidiary arrangements), is to be determined in light of certain criteria spelled out in INFCIRC/153/para.81.

There are five criteria in all. Three of them are straightforward and not very susceptible to manipulation: (1) the form of the nuclear material under control; (2) the characteristics of the state's nuclear fuel cycle; and (3) technical developments in safeguards. The second of these involves the so-called fuel-cycle-oriented approach, which is being studied by the IAEA but involves a number of conceptual and technical problems that have made any significant progress difficult.[39] A fourth criterion, the effectiveness and dependability of the SSAC, has just been discussed, and is clearly dependent upon subjective judgments and interpretation. A fifth criterion, "international interdependence . . . and the extent to which the State's nuclear activities are interrelated with those of other States," while primarily intended to take account of the degree to which countries were dependent on others to maintain an efficient and viable nuclear power program, also embraces the EURATOM case, although it has apparently been less relevant to the establishment of safeguards arrangements in the European Community than have been the provisions related to accounting and control system effectiveness and functional independence of operators.

In conclusion, the actual routine inspection effort undertaken by the agency is governed by a number of considerations. The range of criteria applied in determining an inspection effort is broader than was the case in the earlier safeguards document and contains subjective as well as objective elements. The main importance of these provisions is that they create a basis upon which the agency can exercise some flexibility in the application of safeguards and differentiate among situations in a somewhat standardized manner. However, it also appears that here, as with other measures, the agency is not always able to fully protect its independence when it comes to implementation. States have insisted on specifying the actual routine inspection effort in subsidiary arrangements, although this was not the intent of the safeguards document, thereby placing further real or psychological boundaries on agency discretion. However, as pointed out by a long-term participant in the process, the actual routine inspection effort ultimately agreed to has not been demonstrated to be inadequate, and is even higher than the level of effort the agency actually can support based on resources.[40]

Containment and Surveillance

The basic technique of safeguards is material accountancy (that is, material measurements, records, reports, and verification of the data involved to determine whether material subject to safeguards is accounted for). It was the only technique explicitly covered in INFCIRC/66, although in practice it was often supplemented by containment and surveillance. There was, however, no a priori basis upon which to invoke these methods and, correspondingly, no a priori reason why they could not be invoked. As noted in the preceding chapter, they gained acceptance slowly, and it was only in the mid-1970s that provision for their routine inclusion in safeguards agreements became a firm practice.

INFCIRC/153 specifically incorporates these two safeguards methods as "important complementary measures" to material accountancy, which is designated as a "safeguards measure of fundamental importance."[41] The two measures are explicitly identified in paragraph 74(d) as measures that should be used for the purpose of fulfilling inspection responsibilities.

As a rule, containment consists of measures that take advantage of structural features of nuclear facilities or other locations containing nuclear material that physically restrict or control access to, or movement of, nuclear material. Surveillance entails the use of instrumental or human observation aimed at detecting the movement of nuclear material or tampering with containment measures (such as seals) as well as tamper-detecting features built into unattended instruments. Closed circuit TV and cameras are typical surveillance devices.

The precise meaning of containment and surveillance as "complementary" to material accountancy is somewhat uncertain. It could mean measures to supplement material accountancy when the latter is unable to achieve the desired result of assuring compliance with undertakings or detecting diversion; or it could mean alternatives to material accountancy if the situation seemed appropriate, aside from the effectiveness of accountancy measures.[42]

Containment and surveillance measures are coming to be regarded as increasingly necessary with respect to effective safeguarding of bulk handling facilities. These facilities have large throughputs of accessible nuclear materials, and material accountancy measures cannot provide the desired timeliness of detection of diversion. But they are less amenable to quantification than are measures of material accountancy, and hence it has been difficult to arrive at quantified safeguards outcome statements.[43] Even in the case of INFCIRC/153, with its explicit inclusion of containment and surveillance, the application of such measures is negotiated between the agency and the state and incorporated in the subsidiary arrangements, which once again signals the need, in practice, for mutual agreement to any change. At this point, the two safeguards documents seem to converge on the role and utilization of containment and surveillance measures.

Searching for Objectivity

Beyond the conceptual approaches described above, the NPT safeguards document also sought to minimize subjective judgment and to provide greater precision with respect to the objective of safeguards. As noted earlier, straightforward adoption of INFCIRC/66/REV.2 would have obligated non-nuclear weapon state NPT signatories to accept open-ended statewide safeguards with an inadequate definition of the limits of the measures that might be imposed. INFCIRC/153 sought to remedy this problem by providing an explicit statement of the objective of safeguards.

Accordingly, paragraph 28 asserts that the objective of safeguards is "the timely detection of diversion of significant quantities of nuclear material from peaceful nuclear activities to the manufacture of nuclear weapons or of other nuclear explosive devices *or for purposes unknown*, and deterrence of such diversion by risk of early detection" (emphasis added). Three aspects of this provision deserve comment.

First, this definition of the safeguards objective is somewhat at variance with the general principle stated earlier in the document (153/1.2) that safeguards are applied "for the exclusive purpose of verifying that [source and special fissionable] material is not diverted to nuclear weapons or other nuclear explosive devices." The former (paragraph 28) emphasizes the timely detection of diversion, while the latter (paragraph 1.2) emphasizes verification of nondiversion. This seemingly inconsequential distinction has contributed to some of the controversy in recent years over the extent to which safeguards are effectively fulfilling their nonproliferation role. This subject will be explored in greater detail in chapter 7.

Second, paragraph 28 speaks of diversion "for purposes unknown." The importance of this provision is that it indicates that the agency does not have to seek to determine the use to which diverted material is put, or to prove that diverted material is being used for the manufacture of nuclear weapons or other nuclear explosive devices, but only to conclude that material cannot be accounted for. The burden of accounting for the material falls on the state.

Third, although stated as technical objectives, the key terms of paragraph 28 are still conceptual in nature and need to be translated into practical terms by identifying the appropriate numerical parameters and assigning them quantitative measures.[44] Three such terms appear in the statement of objective, and together with their assigned quantitative indicators are: timely detection (detection time); significant quantity (significant quantity); and risk of early detection (detection probability). *Timely detection* is quantified by relating specified nuclear materials (such as plutonium, high enriched uranium, low enriched uranium, thorium) to the time necessary to convert material in specified form into metallic components suitable for use in a nuclear explosive device. *Significant quantity* is derived by estimating the amount of nuclear material in different forms required for the manufacture of a nuclear explosive device. The values assigned at present by the agency for different materials and times are given in tables 5-2 and 5-3. The agency aims at a *detection probability* of 90–95 percent and a false alarm probability of 5 percent or less, both of which numbers are derived from statistical probability analysis and not an objectively defined rationale. Thus, in these two cases, a satisfactory value remains to be determined.

Another dimension of the effort to provide a technical definition of the objective of safeguards is to be found in paragraph 30 of the safeguards document, which calls for a technical conclusion of the

TABLE 5-2. ESTIMATED MATERIAL CONVERSION TIMES TO FINISHED Pu OR U METAL COMPONENTS

Beginning material form	Conversion time
Pu, HEU, or U-233 metal	Order of days (7–10)
PuO_2, $Pu(NO_3)_4$, or other pure Pu compounds; HEU or U-233 oxide or other pure compounds; MOX or other non-irradiated pure mixtures containing Pu, U[(U-233 + U-235) $\geq$ 20%]; Pu, HEU, and/or U-233 in scrap or other miscellaneous impure compounds	Order of weeks (1–3)[a]
Pu, HEU, or U-233 in irradiated fuel[b]	Order of months (1–3)
U containing < 20% U-235 and U-233; Th	Order of one year

Source: IAEA, *IAEA Safeguards Glossary*, IAEA/SG/INF/1 (Vienna, IAEA, 1980) p. 21.

[a]This range is not determined by any single factor but the pure Pu and U compounds will tend to be at the lower end of the range and the mixtures and scrap at the higher end.

[b]Criteria for establishing the irradiation to which this classification refers are under review.

TABLE 5-3. SIGNIFICANT QUANTITIES

	Material	Significant quantity	Safeguards apply to
Direct-use nuclear material	Pu[a]	8 kg	Total element
	U-233	8 kg	Total isotope
	U[U-235 ≥ 20%]	25 kg	U-235 contained
	Plus rules for mixtures where appropriate		
Indirect-use nuclear material	U[U-235 < 20%][b]	75 kg	U-235 contained
	Th	20 t	Total element
	Plus rules for mixtures where appropriate		

Source: IAEA, *IAEA Safeguards Glossary*, IAEA/SG/INF/1 (Vienna, IAEA, 1980) p. 22.
[a]For Pu containing less than 80% Pu-238.
[b]Including natural and depleted uranium.

IAEA's verification activity in the form of a "statement in respect of each material balance area of the amount of material unaccounted for over a specific period," including a statement of the limits of accuracy of the material balance (thus taking account of problems such as measurement imprecision).

Specific requirements in terms of the quantities of material unaccounted for and detection time have not been established by the board of governors, but the values assigned to the notions of significant quantity and timely detection have been adopted by the board, on a provisional basis, as guidelines for inspection planning and for detection goals. Setting of the goals took place on the recommendation of the Special Advisory Group on Safeguards Implementation (SAGSI) a group of expert advisors established to assist in defining options for dealing with problems associated with translating principles and procedures of the safeguards document into operationally effective and acceptable form. Creation of such a group had been urged by Japan at the time INFCIRC/153 was being formulated, but without immediate result, and it was only several years later that the director general decided on the need for a technical advisory committee.[45]

The introduction of the concept of detection goals has, as we shall see later, generated considerable misunderstanding and controversy. Some have tended to interpret goals as *requirements* and to contend that safeguards are inadequate in light of the agency's inability to meet those goals with respect to certain kinds of facilities. Others have argued that goals are statements of long-term objectives that serve as criteria for planning inspections and for providing a quantitative basis for distributing inspection efforts as well as for judging progress in safeguards effectiveness and as a guide to safeguards research and development efforts.

The latter is the agency's position on the purpose of detection goals. On the one hand, it has adopted inspection goals on a facility-by-facility basis, goals that are technically attainable under state-of-the-art safeguards techniques and methods (the detection goals which may currently be technically unattainable in such cases as large-scale reprocessing facilities). The inspection goals themselves may not be achieved, but the reason for this can be traced to resource deficiencies rather than to technical limitations.[46]

On the other hand, the agency has developed accountancy verification goals that specify the minimum quantity of material that, if diverted from a facility, would be detected by use of material accounting measures, taking account of the need to avoid an unacceptable level of false alarms. They are the best the agency expects to do, given the type and throughput of a facility. The agency also seeks to compensate for deficiencies resulting from primary reliance on material accountancy procedures by supplementing them with containment and surveillance measures.[47]

Whatever goal is being considered—detection, inspection, accountancy verification—in certain cases, such as large plutonium bulk handling facilities, the attainable goal may exceed the significant quantity value identified in table 5-2. This defines a challenge to international safeguards but also to the international community in coming to grips with the dual objective of promoting full access to the peaceful uses of nuclear energy while averting the risk of the proliferation of nuclear weapons. It is a challenge of deciding what role we want and expect safeguards to play in the broader nonproliferation regime, and of defining the criteria that we will use to gauge their effectiveness. While this question could be debated purely in terms of safeguards, it clearly is part of the larger fabric of nonproliferation and it is in that broader context that it ought to be resolved. To limit the debate exclusively to safeguards by implication imposes on safeguards the responsibilities of the total nonproliferation regime. This is both unfair and unrealistic and cannot serve the general interest in limiting the risk of abuse of the peaceful atom.

NOTES

1. Quoted in Mason Willrich, *The Non-Proliferation Treaty* (Charlottesville, Va., Michie Press, 1969) p. 3.

2. Treaty on the Non-Proliferation of Nuclear Weapons, Article III.1. Safeguards apply to declared nuclear material and do not apply to the transfer or receipt of nuclear weapons or devices. Nor are safeguards intended to verify that a state is not making preparations for developing a nuclear explosive device.

3. See David Fischer, "Safeguards Under the Non-Proliferation Treaty," in *Nuclear Safeguards: A Reader*, Report prepared by the Congressional Research Service, Library of Congress, for the Subcommittee on Energy Research and Production. Transmitted to the Committee on Science and Technology, U.S. House of Representatives, 98th Congress, First Session (Washington, D.C., U.S. Government Printing Office, 1983) p. 574. See also, "Background Paper on the Activities of the IAEA Relevant to Article III of the NPT" (Prepared by the Secretariat of the IAEA for the Third Review Conference of the Parties to the Treaty on the Non-Proliferation of Nuclear Weapons) NPT/CONF.III/9 (Vienna, IAEA, June 28, 1985).

4. Hans Grumm, "IAEA Safeguards—Status and Prospects," in *Nuclear Safeguards: A Reader* (see note 3 above) pp. 628–650.

5. IAEA, *The Annual Report for 1984* GC(XXIX)/748. In 1985, the figure for non-nuclear weapons states rose to 887, and in 1986, to 899. *The Annual Report for 1985* GC(XXX)/775 (Vienna, IAEA, July 1986) p. 61) and IAEA, *The Annual Report for 1986*.

6. All figures are derived from IAEA *Annual Reports*.

7. This is not to imply that the Federal Republic of Germany and Japan first signed the NPT and then sought to modify safeguards through the 1970 safeguards committee that drafted INFCIRC/153. Their concerns predated the NPT and led directly to the strategic points and materials-only constraints written into the NPT; they withheld ratification until well after INFCIRC/153 was adopted and certain other issues of concern to them were resolved.

8. For an excellent overview of these developments, see Myron B. Kratzer, "Historical Overview of International Safeguards," IAEA Conference on Nuclear Power Experience, IAEA-CN-42/31 (Vienna, IAEA, September 1982). See also, SIPRI, *Safeguards Against Nuclear Proliferation*, a SIPRI Monograph (Cambridge, Mass., and London, MIT Press, 1975) p. 6.

9. See INFCIRC/66/REV.2, paragraphs 9–14; INFCIRC/153, paragraphs 4 and 5.

10. INFCIRC/153, paragraph 9.

11. Ibid., paragraphs 20–22.

12. Ibid., paragraph 6.

13. Ibid.

14. Ibid., paragraph 29.

15. Ibid., paragraphs 7, 31, 32.

16. Ibid., paragraphs 80, 81.

17. Ibid., paragraph 39, which states that the subsidiary arrangements "shall specify in detail, to the extent necessary to permit the Agency to fulfill its responsibilities under the Agreement in an effective and efficient manner, how the procedures laid down in the Agreement are to be applied."

18. Subsidiary arrangements contain a general part with standardized information regarding inventory, report form, channels of communication, termination of safeguards, etc., and a facility attachments component that includes definitions of material balance areas, key measurement points, frequency and timing of physical inventories and information regarding routine inspections. It is this second component that is kept fully confidential by the inspectorate.

19. For a discussion of this case, see David Fischer and Paul Szasz, *Safeguarding the Atom: A Critical Appraisal* (London and Philadelphia, Taylor and Francis, 1985) pp. 16–17. Indeed, the Pakistani case was relatively straightforward. If safeguards are to evolve through research and development toward improved effectiveness, arrangements will increasingly require modification in ways that are likely to increase the burden on operators. States tend to take the position that current safeguards have been accepted as adequate, and therefore they take a hard line against any intensification of effort for which they concede no justification. The board of governors may well be in a poor position to confront states, insisting that what had been negotiated as acceptable must now be renegotiated. The only way in which such advances can work reasonably well is if states cooperate rather fully and admit that it is worth paying increased costs in one form or another to achieve greater effectiveness.

20. INFCIRC/153, paragraph 116.

21. Ibid., paragraph 6.

22. This should not be taken to imply that safeguards detect diversion by catching diverters red-handed, which they do not. Strategic points are selected in connection with material-balance determination rather than with the physical movement of material. However, since those points are selected with a view to the ease of determining the movement and transfer of nuclear material, they often relate to places where material is relatively accessible.

23. INFCIRC/153, paragraph 76c.

24. Ibid., paragraphs 73, 77.

25. The problems mentioned earlier in relation to Pakistan, admittedly involving a non-NPT state operating under a safeguards document that did not make explicit provision for upgrading, are at least one measure of the potential difficulties that may arise.

26. INFCIRC/153, paragraph 7.

27. Ibid., paragraph 32.

28. While the agency counts on effective SSACs to facilitate the application of safeguards, its information is not confined to reports made by the state to the IAEA. At the facility, for example, all of the operating records are available to the inspector, thus taking his information beyond what may have been provided in the formal report. For a description of SSACs, see IAEA, *Guidelines for States' Systems of Accounting for and Control of Nuclear Materials* IAEA/SG/INF/2 (Vienna, IAEA, 1980).

29. INFCIRC/153, paragraph 7.

30. Ibid., paragraph 81. The term "state" should be read to include such regional arrangements as EURATOM.

31. At the final 1967 session of the Eighteen Nation Disarmament Conference, the Soviet delegate insisted that there should be only "a single system of control for all non-nuclear states so that no non-nuclear country would have special privileges." UN Document ENDC/PV.356, paragraph 15, December 14, 1967. (ENDC was the forum in which the NPT was negotiated.)

32. For a review of this subject, see Paul Szasz, "International Atomic Energy Safeguards," in Mason Willrich (ed.) *International Safeguards and Nuclear Industry* (Baltimore, Md., The Johns Hopkins University Press, 1973) pp. 137–138.

33. "Nuclear Safeguards," *Adelphi Papers*, No. 86 (March 1972) p. 14. It must be added that the concern expressed here was primarily aimed at ensuring that Japan was not differentially treated from the EURATOM states rather than at establishing a general principle applicable in each and every case. The Soviet Union, regarding EURATOM as little more than a tool of NATO, shared the Japanese view.

34. INFCIRC/66/REV.2, paragraph 47.

35. Ibid., paragraph 57, and in both of the extensions (REV.1 and REV.2) to reprocessing and fuel fabrication facilities.

36. Ibid., paragraph 58.

37. INFCIRC/153, paragraph 80.

38. Ibid., paragraph 79.

39. For a discussion of this issue, see chapter 7 herein and Andre Petit, "De la necessité d'approfondir le consensus international quant aux implications pratiques des concepts de base des garanties internationales," reproduced in Aspen Institute for Humanistic Studies, *Proliferation, Politics and the IAEA: The Issue of Nuclear Safeguards*, Part II-Berlin (Queenstown, Md., Aspen Institute for Humanistic Studies, 1985) Appendix II.

40. David Fischer in David Fischer and Paul Szasz, *Safeguarding the Atom: A Critical Appraisal* (see note 19 above) p. 62.

41. INFCIRC/153, paragraph 29.

42. See Myron B. Kratzer, "New Trends in Safeguards," Proceedings of the 26th Annual Meeting of the Institute of Nuclear Materials Management, vol. XIV no. 3 (Albuquerque, N.M., July 21–24, 1985) pp. 10–15.

43. There are significant differences in the safeguards community over the role containment and surveillance can play. The argument against considering containment and surveillance as anything more than a supplement to material accounting is that INFCIRC/153 paragraph 30 states that the product of safeguards is the material balance statement and that containment and surveillance cannot provide such a statement. It is also argued that containment and surveillance could not detect a diversion apart from a determination by accounting that material was missing and could not be otherwise accounted for. In this view, the most containment and surveillance can do is to support accounting by providing confidence that flows and inventories were counted and counted only once; and to preserve measurements, as with material in sealed containers. But it cannot in this view be an alternative to accounting and cannot provide precision of findings or timeliness of detection beyond what accounting can provide.

44. The author is indebted to James de Montmollin and Eugene Weinstock for the many discussions they held with him, even if their advice was not always adopted. For an insight into some of the problems covered here, see their paper, "The Goals of Measurement Systems for International Safeguards," American Nuclear Society/Institute of Nuclear Materials Management Topic Conference on Measurement Technology for Safeguards and Materials Control, Kiawah Island (ANS/INMM, November 26, 1979; mimeo).

45. See Ryukichi Imai, "Nuclear Safeguards," *Adelphi Papers* (see note 33 above) p. 7; and David Fischer and Paul Szasz, *Safeguarding the Atom: A Critical Appraisal* (see note 19 above) pp. 67–68.

46. Inspection goals are essentially based on one-explosive quantities or the feasible limit of measurement accuracy, whichever is larger.

47. As noted earlier (see note 43 above), not all participants in the safeguards community agree with the notion that containment and surveillance can play a compensatory role although they do not deny that they can modestly support material accountancy.

Chapter 6

NUCLEAR POLICIES IN TRANSITION

LANDMARK EVENTS

The Treaty on the Non-Proliferation of Nuclear Weapons (NPT) came into force in 1970. The basic safeguards arrangements it called for were agreed on in 1971. Consequently, there was every reason to anticipate a sustained period of stable and increasing international nuclear cooperation. But as discussed briefly in chapter 1, this was not to be the case. The watershed year 1974, just after the twentieth anniversary of Atoms for Peace, marked the opening for one of the most unsettled and controversial periods of the nuclear age.[1] Many of the issues raised during this era remain unsolved at this writing. In some cases, they are a continuing irritant in relations between nuclear suppliers and recipients, or between nuclear weapon and non-nuclear weapon states. In other cases, however, changing political, economic, and energy circumstances have reduced the importance of some issues, at least for the present. In one way or another, most of the events and problems that began to emerge in 1974 also have had an impact on the International Atomic Energy Agency (IAEA).

While an in-depth discussion of all facets of nonproliferation lies beyond our purpose, a review of key events, developments, and policies since 1974 is important to an understanding of some of the difficulties confronting the IAEA. Several were mentioned in chapter 1 in the discussion of the third phase of the evolution of the nonproliferation regime. Accordingly, we shall endeavor here to avoid repetition as much as possible.

India's Nuclear Test

One of the most important events in the history of nonproliferation was India's test of a "peaceful nuclear explosive" on May 18, 1974. The incident focused attention on all aspects of the proliferation problem and stimulated reactions and responses that unfolded over many years and whose impact still reverberates throughout the nuclear world. It remains the first and only acknowledged instance in which external nuclear assistance explicitly made available for peaceful purposes was used to assist in the provision of material for a nuclear explosive.[2] Beyond all doubt, it was a severe blow to the hopes for nuclear stability that had been building around the NPT and the nonproliferation regime, and the hope that an international nonproliferation ethic would attain universal acceptance. In addition, it galvanized new forces, particularly in the United States, that were gathering on the fringes of the nuclear power camp and questioning both the wisdom of promoting domestic use of this awesome technology and the efficacy of a policy of controlled nuclear cooperation abroad.

Severe and far-reaching reactions were directed more toward international safeguards and the nonproliferation regime than toward India itself, a paradoxical response in view of the fact that India had adopted only minimal nonproliferation obligations, inscribed in what turned out to be rather loosely written agreements with its suppliers, had not joined the broader nonproliferation regime, and had separated the plutonium used in the explosive device in an indigenously built, unsafeguarded reprocessing plant. Yet it was the regime as a whole that was to bear the burden of India's action.[3] India was not a party to the NPT and had not agreed to the treaty's prohibition against the manufacture or acquisition by other means of nuclear weapons or other nuclear explosive devices. Nor did the treaty enjoy such broad adherence and support as to express a customary rule of international law binding on all nations.[4] Furthermore, as a party to the Partial Test Ban Treaty of 1963, India had conducted its nuclear test underground, consistent with provisions of that treaty. Thus, no formal international regime commitments had been breached, even if India's action had contravened the understandings and expectations of others and had acted in a manner inconsistent with the spirit of the nonproliferation regime.

On the other hand, there is the argument that India willfully misinterpreted bilateral agreements that also are a part of the regime.

The reactor that produced the plutonium used in the 1974 test had been provided by Canada under an agreement that it be used for exclusively peaceful purposes, an agreement without an explicit exclusion of "peaceful nuclear explosives" and without safeguards to verify use of the reactor. The United States had supplied heavy water to India for use in the reactor under similar ambiguous conditions. From the early 1970s, the Canadian government had tried unsuccessfully to reach agreement with India on an interpretation of peaceful uses that would preclude any nuclear explosive activity of any kind—a forcefully stated and *unambiguous* Canadian position. Immediately after the test, Canada interrupted supplies of nuclear material and equipment to India. Two years later, Canada canceled the bilateral cooperation agreement, having failed in its effort to reach an acceptable agreement with India that would ensure that Canadian nuclear exports would not again be used for nuclear explosive purposes.[5]

The United States, for its part, imposed a temporary embargo on nuclear supplies to India, but lifted it upon receiving India's assurance that the U.S. nuclear fuel supplies for the Tarapur power reactors (the main component of U.S.-Indian cooperation) would not be used to produce nuclear explosives of any kind. Several years later, new U.S. nonproliferation legislation would lead to an impasse over nuclear cooperation with India resulting in termination of U.S. supply of nuclear fuel. Thus, India's test highlighted deficiencies in agreements for cooperation negotiated by two supplier states, and reflected the imprecision surrounding early definitions of peaceful nuclear activity. It also underscored the danger of plutonium separation technology and the relative ease with which peaceful plutonium could be turned to explosives.

Nevertheless, India's explosion raised for some a question whether the NPT and the regime as it stood were sufficient to prevent more states from acquiring nuclear weapons or other nuclear explosives. This concern was driven not only by the Indian test, which demonstrated both the ability of Third World nations to master sophisticated technologies, and the risk of the abuse of external assistance, but by other events as well. Two of them—one involving projected transfers of sensitive fuel cycle technology by other suppliers to Third World nations, and the other a proposed sale of nuclear reactors by the United States to countries in the Middle East—deserve special comment.

Transfer of Sensitive Nuclear Technology to the Third World

The projected transfer to several Third World states of facilities and technology that could provide direct access to weapons-usable material intensified the already existing concern about the adequacy and effectiveness of the nonproliferation regime. The potential recipients—South Korea, Taiwan, Pakistan, and Brazil—all lacked nuclear power programs that were sufficiently large or technically advanced to offer any economic or technological justification for the acquisition of advanced technologies such as reprocessing, or for access to separated plutonium. Moreover, with the exception of Brazil, they all were located in unstable regions, were politically and militarily vulnerable, and were suspected of having an interest in nuclear weapons or at least a nuclear weapons option—a suspicion ultimately confirmed in more than one case and still very much alive with respect to Pakistan.[6]

Unlike the Indian case, there was no question about restrictions on the use or application of safeguards in these countries. The intended supplier, France, had included clear limitations on use and imposed rigorous safeguards. Rather, the issue was whether such facilities should be exported in the first place, and whether the safeguards would be sufficient not only to detect, but to prevent, any diversion that might be attempted. These reasonable concerns signaled the first of many occasions when differences would arise over the capability of safeguards to provide adequate protection against the risk of diversion, and when confusion about the scope and purpose of international safeguards would suffice.[7]

By 1976, U.S. diplomatic efforts had brought about the cancellation of Taiwanese and South Korean plans to acquire reprocessing facilities. Similar efforts in Pakistan were less successful, even in the face of the U.S. cancellation of military and economic assistance to Pakistan for several years (which was, however, partly neutralized by new assistance from other Western states). Moreover, neither Canadian cancellation of nuclear cooperation with Pakistan, affecting Pakistan's only nuclear power reactor, nor eventual French suspension of assistance to Pakistan on the reprocessing plant, was sufficient to deter Pakistan from continuing efforts to acquire a reprocessing, and later an enrichment, capability.[8] Nor was the United States able to influence the proposed transfer of reprocessing and enrichment technology under very rigorous safeguards by the Federal Republic of Germany to Brazil, a non-NPT state.[9] All of the events only served to deepen concerns abut nonproliferation, the adequacy of the regime, and international security.

Proposed U.S. Reactor Sales to Egypt and Israel

The U.S. proposal to sell nuclear power reactors to Egypt and Israel arose out of a peace mission by President Richard Nixon to the Middle East just weeks after the Indian explosion. The announcement of the offer, which apparently was made with little, if any, prior planning, created a sense of anxiety in some quarters in the United States, particularly in Congress.[10] With the Indian incident fresh in mind, questions were raised as to whether bilateral agreements guaranteeing peaceful use could be relied upon over the longer term, especially in circumstances involving states with a long history of conflict and unstable relations. There was also doubt about whether the reactors might not eventually be used to help acquire nuclear weapons. These concerns were heightened because neither Egypt nor Israel, like India, were parties to the NPT and because neither treaty adherence nor full-scope safeguards were being made a condition of cooperation. The reliability of international safeguards arrangements was also called into question in ways that suggested some confusion over the role and purpose of nonproliferation safeguards. Thus, both the risk of unilateral abrogation of commitments and the absence of punitive measures to deal with such an incident often were lumped together with questions about safeguards adequacy. More discerning questions about safeguards effectiveness, particularly in relation to sensitive technology and facilities, were to emerge later in relation to the French and German fuel cycle sales just noted.

Coming as it did in the wake of Vietnam and the midst of the Watergate crisis, both of which had severely shaken confidence in the executive branch of government, President Nixon's nuclear initiative impelled Congress to seek to strengthen its role in the nuclear decision-making process. Accordingly, Congress amended the Atomic Energy Act of 1954 to require that any proposed international agreements for peaceful nuclear cooperation involving reactors larger than five megawatts be submitted to Congress for sixty days of continuous session during which time Congress could disapprove the agreement by concurrent resolution. The measure was signed into law by President Gerald Ford in October 1974.[11] This was to be only the first of several steps designed by Congress to strengthen its role in nuclear nonproliferation policy.[12] The important point for our purpose is that Congress became restive with routine negotiation of agreements and despite the existence of the Joint Committee on Atomic Energy sought to open these to broader scrutiny.[13]

THE DANGER OF HIGH-RISK EXPORTS

In discussing nonproliferation, emphasis inevitably must be given to the importance of the United States. Almost every significant nonproliferation initiative has originated with the United States, beginning with the Baruch Plan and including Atoms for Peace, the instituting of the principle of verification safeguards, the creation of the IAEA, and promotion of the NPT. The U.S. role continued to be central in the aftermath of the Indian explosion and the revelation of the agreements for the transfer of sensitive nuclear technologies to developing countries with limited nuclear power programs.

The pattern of events described above did not conform with the United States' initial expectations in promoting the peaceful atom: it was different, it was worrisome, and it was deemed to require some kind of response if further erosion of the status quo was to be avoided. The development of nuclear power under the nonproliferation regime brought with it a conventional wisdom on nuclear fuel cycle development and a set of expectations regarding the pace and character of evolving industrial nuclear programs. It was widely expected that at a certain threshold point in nuclear power development, reprocessing of discharged fuel to recover its residual energy value in the form of plutonium and unburnt uranium would be appropriate and necessary, as would the development of advanced reactors that would yield greater efficiencies in nuclear use, in particular breeder reactors. This expectation was predicated largely on a belief in the limited availability of accessible and economic uranium that could be exhausted by a rapidly expanding demand for nuclear energy. These expectations were based on widely shared assumptions and projections about rising demand for electricity.

Although not stated explicitly, it was logically assumed that the development of advanced fuel cycle activities would initially take place in the leading nuclear industrial states and not in developing countries that, to the extent they were involved at all, were for the most part only in the beginning stages of nuclear development. Indeed, whatever ultimate expectations the United States and other suppliers may have had during the 1950s and 1960s about the geographic reach of advanced nuclear technology, it is likely that their field of vision (at least for extensive applications of nuclear power) was largely bounded by the industrial world and perhaps several unusually advanced (scientifically) developing countries such as India. It is probable that while anticipating progressive expansion of nuclear energy into all regions of the world, there was little expectation of near-term demands (except possibly from India) for the more sophisticated or dangerous nuclear technologies from the developing countries (because they could serve no logical use at that stage of the countries' nuclear development), and a belief that there was a relatively good fit between the controls being deployed and the reliability and stability of the most likely importing countries.

This is not to say that an eventual broad dissemination of nuclear power related activity was not anticipated. Quite the contrary. In 1973, the IAEA had conducted a market survey of nuclear power estimates in fourteen developing countries. That survey suggested that between 52,000 and 62,000 MW(e) of nuclear plant capacity might be put into operation in those countries by 1990.[14]

Though excessively optimistic, and subsequently challenged by other analyses on economic, technological, and industrial infrastructure grounds,[15] the IAEA survey nevertheless reflected the accepted view that a number of developing countries would soon begin to enter the nuclear power age, if only modestly. What was not contemplated was that some of these countries would be able to use the emerging liberal trade market to acquire the means to produce weapons-usable material without regard to the logic of that effort to their plans for nuclear power development.

The partial unraveling of what had been up to then a rather disciplined and well-controlled system of international nuclear cooperation was due to several factors. One was the change in the structure of the nuclear marketplace that was marked by the appearance of other nuclear suppliers competing with the United States; these suppliers, while not indifferent to nonproliferation, nevertheless had different views on what constituted necessary and adequate controls. They were sometimes guided more by commercial considerations than by hypotheses on long-term risk of nuclear technology dissemination. Some tended to take as sufficient the safeguarded peaceful assurances that they received regarding intended use of transferred technologies or facilities, whereas in similar circumstances, the United States might be less prepared to settle for such assurances. This factor was only partly subject to U.S. control.[16]

A second and related factor that was very much subject to U.S. control was the decline of U.S. predominance in enriched uranium supply. The United States enjoyed a virtual monopoly in furnishing power reactor fuel as of the early 1970s. But a decision by the Nixon administration to put enrichment services on a commercial footing and to turn enrichment production over to the private sector undercut the ability of the United States to maintain its market position. In 1972, the United States changed its conditions for contracting enrichment services from a system of "requirements" to one of "fixed commitments," which also involved substantial prepayment by customers. This had a very unsettling effect on the market and also generated inordinate demand as domestic and foreign customers sought to ensure themselves adequate supplies. The U.S. enrichment capacity remained constant and became progressively saturated as the internal battle to achieve privatization proceeded. Under law, it was prohibited to undertake enrichment contract services beyond a stipulated capacity. As a result, the United States no longer was in a position to enter new contracts for enrichment by 1974.

In July 1974, the United States announced that it was suspending the signing of any new enrichment contracts and was closing its order books. Among other things, this left a large number of potential customers in limbo, created a crisis of confidence in the reliability of the United States as a nuclear fuel supplier, and served to affirm and consolidate plans elsewhere to develop enrichment capacity. The net result was that the United States lost the predominance it had enjoyed in the nuclear fuel market, and faced the spread of technologies that it had labored hard to keep limited. Further, important elements of the structure upon which nonproliferation was based had been weakened. It was a matter of domestic political priorities getting in the way of international interests and, in retrospect, a classic case of shooting oneself in the foot.[17]

Contrary to some of the views expressed during the course of the nonproliferation debate of the last ten years, U.S. export and cooperation policy always had dealt circumspectly with the issue of dangerous fuel cycle technologies. The United States has never supplied uranium enrichment technology, although it did propose sharing it with an appropriate multinational venture in the early 1970s when confronted with the reality that some of its European allies were planning to proceed with their own enrichment plans.[18] When the members of the URENCO gas centrifuge enrichment consortium (the United Kingdom, the Netherlands, and the Federal Republic of Germany) first began discussing a joint arrangement in the early 1960s, however, the United States had prevailed on them to place their activities under strict secrecy to avoid the risk of unnecessary dissemination of a technology bearing high risk for nuclear proliferation.

As for reprocessing and plutonium fabrication technology, U.S. bilateral agreements for cooperation had provisions contemplating eventual reprocessing under specified terms, conditions, and limitations. It was France, not the United States, that first published basic technical information on reprocessing technology. U.S. information sharing was in a real sense provoked by the French initiative at the 1955 Geneva Conference on the Peaceful Uses of Atomic Energy, after which the United States began to publish technical reprocessing information. It did not, however, share industrial know-how or transfer hardware or facilities. The only instance of actual sharing related to the joint EUROCHEMIC venture (a creation of the European Nuclear Energy Agency, the nuclear arm of the Organization of Economic Cooperation and Development), which was modestly assisted in the hope that any commercial reprocessing activity that might develop on the European continent would be multinationalized and thereby avoid wide dispersion of nationally owned and controlled facilities.

In 1972, after the NPT had come into force, but prior to the Indian test and the revelation of France's contracts with South Korea and Pakistan, the United States revised its internal rules to tighten the conditions under which any private U.S. individual or concern could assist in the development of reprocessing capability abroad. Any such assistance was made contingent on explicit authorization by the executive branch (at first the chairman of Atomic Energy Commission, later the administrator of the Energy Research and Development Administration, and now the secretary of energy), and criteria for evaluating whether to grant such approval were established, namely the NPT status of the potential recipient and whether the facility would be under multinational auspices.[19] The intent was to hold out the possibility for U.S. cooperation in reprocessing as leverage to encourage countries to join the NPT, and to encourage others to seriously consider joint ventures in reprocessing in lieu of establishing independent facilities. Thus, serious and sustained efforts to control the risk of proliferation while advancing the cause of peaceful atomic energy lay in the background of the unsettling events of 1974 and 1975.

THE DILEMMA OF DEFINING PROLIFERATION

Since 1974, discussion, debate, and policies on nonproliferation have been dominated by two interdependent themes—the meaning of proliferation and the extent to which it is prudent to rely on pledges verified by international safeguards for achieving the objective of nonproliferation. To a large extent, this has been a debate between the United States and other countries, but within the United States, it has sometimes been reflected in differing views of the executive and legislative branches of government.

The NPT, which is the juridical and political centerpiece of the nonproliferation regime, defines proliferation in terms of possession of nuclear weapons or nuclear explosive devices, and obliges its signatories to neither transfer nor receive, manufacture, or acquire such devices. The scope of the undertaking that non-nuclear weapon state parties understood themselves to be adopting is reflected in a note transmitted to the United States by the Federal Republic of Germany when the latter signed the NPT. The Bonn government emphasized its understanding that beyond preventing acquisition of nuclear weapons or other nuclear explosive devices, in no case would the provisions of the treaty "lead to restricting the use of nuclear energy for other purposes by non-nuclear weapon states." In particular it said:

> . . . no nuclear activities in the fields of research, development, manufacture or use for peaceful purposes are prohibited nor can the transfer of information, materials and equipment be denied to non-nuclear weapon states merely on the basis of allegations that such activities or transfers could be used for the manufacture of nuclear weapons or other nuclear explosive devices.[20]

There is nothing in the record to indicate a contrary understanding on the part of the United States at the time.

Nevertheless, the meaning of proliferation has been changing in the public debate ever since 1974. Critics concerned about the implications of the trend of spreading nuclear technology have sought to focus public attention on the risk that under safeguards, increasing numbers of countries with nuclear power programs would come close to nuclear explosives without actually violating their commitments not to produce them, especially if they legitimately could have separated plutonium or the facilities to produce it. They have argued that traditional assumptions of Atoms for Peace needed to be reconsidered if a situation described by Albert Wohlstetter as "a legitimate—but Damoclean—'overhang' " was to be avoided. Failure to do so, it was argued, would lead to living in a "nuclear armed crowd."[21]

Such analyses, which already had the sympathetic ear of some officials in the Nuclear Regulatory Commission and the Arms Control and Disarmament Agency, gave a new argument to public interest groups already wary of nuclear power to urge a thorough review of the United States' nuclear cooperation and control policies. These pressures, coupled with the shock of the Indian nuclear test and the other events of the mid-1970s, provoked Congress into a series of hearings aimed at evaluating the soundness of the rules and arrangements for nuclear cooperation then in effect and at tightening related policies and procedures. This evoked the second theme of the post-1974 era, the extent to which international nuclear cooperation—consistent with the avoidance of proliferation—could reasonably depend on a safeguards-based nonproliferation regime to provide adequate nonproliferation assurance, or whether other measures—including more restrictions on trade and cooperation—were needed.

The Legislative Response

The evolution of the definition of proliferation, at least from the congressional point of view, can be traced through legislation culminating in the Nuclear Non-Proliferation Act of 1978 (NNPA). Among the more important predecessors were the Symington and Glenn amendments to the Foreign Assistance Act of 1961. Both measures defined activities deemed to warrant punitive measures on the part of the United States. The amendments implied an expanded definition of proliferation. Under their terms, proliferation included not only the detonation of a nuclear explosive device, but also the acquisition of the capability to produce weapons-usable material even though such activities might also serve peaceful nuclear ends.

In 1976, the Symington amendment[22] provided for a cutoff of economic and military assistance to any country that imported or exported reprocessing or enrichment materials, equipment, or technology *unless* it agreed to place all such items under multilateral auspices and management when available, and the recipient accepted IAEA safeguards on all of its nuclear fuel and facilities, that is, full-scope safeguards. The Glenn amendment of 1977[23] reaffirmed the provisions of the Symington amendment with regard to uranium enrichment, but treated reprocessing differently. Any country that delivered or received reprocessing equipment, materials, or technology was subject to a cutoff of economic and military assistance *regardless* of whether or not it accepted full-scope safeguards and placed the relevant items under available multilateral auspices. By implication, the very act of engaging in any reprocessing-related activity was considered so nefarious as to be tantamount to proliferation, although the Glenn amendment did not go quite that far. Implicitly, international safeguards were considered to be inadequate to deal with this technology and its product.

Both amendments provided for presidential waiver of the cutoff. Under the tougher Symington amendment, the president could waive cutoff if he certified in writing to the speaker of the House of Representatives and to the Senate Foreign Relations Committee that the termination of aid "would have serious adverse effect on vital United States interests" and that "he has received reliable assurances that the country in question will not acquire or develop nuclear weapons. . . ." Waiver under the Glenn amendment only requires that the president determine and certify in writing that termination "would be seriously prejudicial to achievement of United States nonproliferation objectives or would otherwise seriously jeopardize the common defense and security." The gravity of the behavior was not altered by these provisions, but they enabled the president to evaluate behavior in terms of nonproliferation objectives and national interest.

U.S. assistance to Pakistan was suspended in September 1977, under the terms of the Glenn amendment because of Pakistan's continued effort to acquire a reprocessing plant under its agreement with France. French suspension of cooperation on reprocessing with Pakistan in the summer of 1978 resulted in restoration of U.S. aid. In the spring of 1979, the United States once again terminated assistance, this time formally under the Symington amendment because of Pakistan's persistence in seeking to complete an indigenous enrichment facility at Kahuta.

The subsequent Soviet intervention in Afghanistan in December 1979 led President Jimmy Carter to offer to resume and even increase military assistance to Pakistan as a demonstration of U.S. commitment to Pakistan's national security and its opposition to the Soviet presence in Afghanistan even though Pakistani nuclear activity made it difficult, if not impossible, to exercise a presidential waiver. The question of the scope and magnitude of assistance dragged on into the early months of the Reagan administration and finally was settled in the form of a proposed $3.2 billion aid package, including state-of-the-art F-16 aircraft, over a period of five years. In arguing for the proposed aid package before Congress, administration spokesmen indicated that Pakistan had pledged not to develop nuclear weapons,[24] and that the administration had made unequivocally clear to Pakistan that aid would be terminated should that country conduct a nuclear test.[25] In authorizing implementation of the aid program, Congress further tightened the Symington and Glenn amendments by inserting a provision requiring termination of aid if Pakistan received or transferred a nuclear explosive device or conducted a nuclear test, and by adding a congressional veto by concurrent resolution of a presidential decision to continue aid in the context of a violation of nonproliferation conditions.[26]

The Nuclear Non-Proliferation Act of 1978 (NNPA) is the most comprehensive and important legislation on peaceful nuclear cooperation and nonproliferation since the Atomic Energy Act of 1954, which initially opened the door to cooperation and exports under Atoms for Peace. It contains many of the conditions and criteria that govern U.S. peaceful nuclear cooperation and exports and establishes the statutory framework for executive branch policies. The NNPA reaffirmed U.S. commitment "to a strengthened and more effective International Atomic Energy Agency and to a comprehensive safeguards system administered by the Agency," and called for international efforts to provide the necessary funds, technical resources, and other support necessary for effective implementation of a strengthened safeguards program.[27]

But it also reflected a congressional view that a safeguards-based nonproliferation regime alone was not sufficient to prevent proliferation, even with respect to NPT parties. To avoid deterioration of the integrity of the regime, additional measures were needed to avert the risk that sensitive nuclear materials (such as plutonium and highly enriched uranium) and the technologies for their production would come into widespread use. The title of the NNPA, "An Act to provide for more efficient and effective control over the proliferation of nuclear weapon *capability*," [emphasis supplied] underlines this concern as does the preamble, which states that:

> The Congress finds . . . that the proliferation of nuclear explosive devices or of the direct capability to manufacture or otherwise acquire such devices poses a grave threat to the security of the United States and to continued international progress toward world peace . . . [and therefore adopts a policy to establish] more effective international controls over the transfer and use of nuclear materials and equipment and nuclear technology. . . .[28]

The NNPA thus continued the trend started with the Symington and Glenn amendments of expanding the definition of proliferation and of implicitly questioning the effectiveness of safeguards in cases involving sensitive nuclear materials and the facilities to produce them, even while recognizing the general importance of safeguards to nonproliferation.

Pursuant to these concerns, the NNPA stipulated a number of additional measures aimed at constraining the spread of sensitive nuclear capabilities. For one thing, the act created a new category of information, "sensitive nuclear technology," defined as information not available to the public and "important to the design, construction, fabrication, operation or maintenance" of enrichment, reprocessing, or heavy water production facilities.[29] Cooperation with respect to, and transfer or export of, such technology can take place only under rigorous conditions established for nuclear exports and cooperation in general.[30] These include a requirement that IAEA safeguards will apply to all activities in the country (at least after two years); guarantees that items will not be used for any nuclear explosive purpose whatsoever, including research, and that they will be afforded adequate physical protection; and that no retransfer of supplied technology or transfer or alteration (that is, reprocessing) of material produced as a result of its use will take place without prior U.S. approval.

Beyond this, the act discourages cooperation or transfers involving sensitive nuclear technology, although in a less than forceful manner. For example, the president is enjoined to *seek* international agreement that enrichment, reprocessing, and fabrication of fuels using weapons-usable material should be carried out only in facilities "under international auspices" with strict limits on non-nuclear weapon state access to sensitive nuclear technology, although a half-dozen existing facilities were grandfathered.[31] The president is also mandated to *seek* to develop effective fuel assurance arrangements. However, the legislation stipulates[32] that the president should *seek* to ensure that fuel assurance benefits are available only to non-nuclear weapon states that "do not establish any new enrichment or reprocessing facilities under their de facto or de jure control, and place any such existing facilities under effective international auspices and inspection." (This provision related to efforts to establish an International Nuclear Fuel Authority, which never materialized.)

In addition to the preceding provisions, the NNPA broadened the definition of circumstances that could result in termination of U.S. nuclear supply. Included were actions that would be expected to evoke sanctions (such as detonating a nuclear device or terminating, abrogating, or materially violating IAEA safeguards) as well as activities having direct significance for the manufacture or acquisition of nuclear explosive devices (even if not intended for that purpose). Also, countries became vulnerable to sanctions if they entered into an agreement to transfer reprocessing equipment, material, or technology to the sovereign control of a non-nuclear weapon state unless the agreement were pursuant to an arrangement to which the United States subscribed.[33]

Upping the Ante for Reprocessing

Still another reflection of the broad meaning of proliferation and doubts about safeguards effectiveness for sensitive nuclear activities and materials can be seen in the NNPA conditions that mandated U.S. approval of the reprocessing of materials subject to its control. Congress provided that approval of a request to reprocess U.S. origin fuel or fuel irradiated in U.S.-supplied reactors (a so-called subsequent arrangement) would require a finding by the secretary of energy (with concurrence of the secretary of state) that the activity "will not result in a significant increase of the risk of proliferation beyond that which exists at the time that approval is requested." Moreover, the act specifies that:

> Among all the factors in making this judgment, foremost consideration will be given to whether or not the reprocessing . . . will take place under conditions that will ensure *timely warning* to the United States of any diversion well in advance of the time at which the non-nuclear-weapon-state could transform the diverted material into a nuclear explosive device.[34] [emphasis added]

The criterion of timely warning generated considerable debate. Some members of Congress argued that it should be the only criterion for approving a request. But this was rejected in favor of the language finally adopted. In implementing this provision, both the Carter and the Reagan administrations in practice assessed the risk of proliferation not only in terms of the timeliness of warning of diversion that safeguards could provide, but also by the NPT status and nonproliferation commitment of the state or states making the request, the technical and institutional arrangements associated with the reprocessing facility (including safeguards and multinational participation), the adequacy of physical protection and safeguards arrangements for separated plutonium, and the right of the United States to approve the return or retransfer of such plutonium.

The Carter administration established additional guidelines for approving the reprocessing of U.S. origin fuel that were based on the principle of case-by-case review. Initially, requests were to be considered for approval only if they involved a clear need (for example, resulting from congestion in spent fuel storage facilities).

However, in order to be politically responsive, particularly to Japanese concerns, another criterion was introduced at a very early stage: the possibility of approving reprocessing transfers to permit the fulfillment of contractual arrangements antedating the new U.S. policy toward processing. Japan had a number of such contracts with France. As a general rule, all approvals were to be subject to the condition that the requesting state was cooperating in exploring alternative methods of spent fuel disposal and that the approval would further U.S. nonproliferation objectives. Japan's operation of the Tokai-Mura reprocessing plant to support both experimental work and safeguards development fulfilled this requirement.[35] The Carter administration generally exercised its prerogatives in pragmatic fashion, thereby minimizing controversy that might otherwise have surfaced; but the administration's general posture was to try to discourage reprocessing to the extent feasible. Whether Congress regarded this pragmatism as fulfilling its objectives is another question.

The Reagan administration altered the policy of case-by-case review to allow for programmatic (that is, long-term) reprocessing approvals in cases involving countries with advanced nuclear programs that do not pose any proliferation threat. The proposed arrangement, intended to be limited to Japan and the EURATOM countries, but more recently enlarged to include Switzerland, was to be implemented in the context of renegotiated agreements for cooperation and a broad understanding on enhanced international cooperation to improve the nonproliferation regime and international safeguards.[36] To a large extent, this policy remains on paper; the relevant negotiations have not been consummated. Nevertheless, announcement of the Reagan approach evoked congressional reaction that took the form of proposed new legislation designed to further amend and tighten U.S. nuclear export and cooperation policy. For many reasons, none of this has come to pass. But the very fact that new legislation has been considered gives evidence of continued congressional interest in nonproliferation policy and strongly suggests that Congress would be quick to take up nonproliferation policy questions if in its view the objectives of U.S. policy as reflected in legislation (assuming Congress has not chosen to amend it) were put at risk.[37]

This brief review of legislative activism shows that during the past decade Congress has been aggressive in the nonproliferation arena. The principal effort has been to tighten export controls, constrain dissemination of certain technologies and facilities, and (through the pressure of legislative oversight) protect against what some felt might be a too liberal administration approach to international nuclear cooperation and exports. Individually and collectively, these measures reflect a congressional uneasiness about the effectiveness of international safeguards as a means of preventing further proliferation. What is striking in all of this is the scant attention given by some legislators to the legitimate aspirations, expectations, and concerns of recipient nations both in the realm of energy security and nuclear development. Moreover, there is little evidence that they appreciate that the stability of the nonproliferation regime would be seriously undermined if nuclear recipients lose confidence in its basic principles.

There is in all of this, of course, an element of action and reaction. The promotion of nuclear commerce generates a certain instability if technology is perceived as getting ahead of social controls that enjoy public confidence. Thoughtful people raised legitimate questions about whether the course and pace of nuclear development were consistent with the level of control perceived to be needed. Fearing not, corrective measures were sought and introduced in an effort to redress the balance and to bring things back to equilibrium. The strong legislative pressures to curtail old patterns in the nuclear community served notice of what could happen if confidence in effective controls waned. However, these measures sometimes ran roughshod over legitimate peaceful interests and somewhat undermined the very institution created to serve the interest of the safe development of the peaceful atom, the International Atomic Energy Agency.

The Executive Response

There was also policy reaction to nuclear developments with far-reaching consequences in the executive branch. Indeed, in a number of instances, the Executive was moving forward more quickly than Congress in an effort to shore up nonproliferation. Despite a deeper understanding of the nature and purpose of safeguards, and widespread belief in their capacity to detect diversions, executive branch confidence in the adequacy of international safeguards for sensitive materials was shaken. Diversions might be detected, but revocations of commitments could not be prevented. Access to weapons-usable material in politically unstable environments therefore presented a problem for the regime, and safeguards alone did not seem to provide an acceptable answer.

Nuclear Supplier Guidelines

Executive branch policy in the mid- and late 1970s was aimed at reinforcing the regime and restoring confidence in the concept of controlled nuclear cooperation. It had two central elements. The first was to mobilize the principal exporters (as the United States had done for nonproliferation reasons on other occasions) into a Nuclear Suppliers Group to achieve agreement on common conditions and rules of the game for nuclear exports. The second was to challenge conventional wisdom regarding the use of plutonium as a nuclear fuel.

A principal objective of the first element was to ensure that safeguards did not become bargaining chips in commercial nuclear competition. The United States favored a requirement that recipient countries accept full-scope safeguards as a condition for any nuclear export. Although backed on this by a number of suppliers (and the condition accepted by all NPT non-nuclear weapon state parties on their own activities), this failed to win French and German support (very largely because of anticipated sales to Argentina, Brazil, and South Africa). So the guidelines provide only that any export of items on an agreed "trigger list" drawn up and occasionally supplemented by the suppliers would have to be placed under IAEA safeguards.[38]

The emphasis of this crucial group of exporting states on international safeguards confirmed that safeguards remained the core of the nonproliferation regime and the sine qua non for international civil nuclear cooperation. Nonetheless, the United States also sought to reduce the pressure on safeguards by pressing for additional technological barriers; this would be accomplished by a mandatory agreement by suppliers that they would not make further transfers of reprocessing or enrichment technology or facilities. This reflected a diminished confidence that safeguards and pledges alone could sustain nonproliferation.

But the U.S. drive for mandatory prohibition on sensitive transfers was rejected by several members of the group as being too sweeping and likely to cause some countries to seek nuclear independence, thereby further diluting any influence suppliers might exercise over national nuclear development. However, the suppliers did agree to exercise restraint in the transfer of sensitive facilities, technology, and weapons-usable materials and to encourage recipients to accept supplier involvement or other appropriate multinational arrangements as an alternative to national facilities.[39] Significantly, the two members of the group least disposed to mandatory restraints on sensitive technology transfers—France and the Federal Republic of Germany—

subsequently independently announced their intention *not* to authorize "until further notice" the export of reprocessing facilities.[40]

Additionally, the guidelines eventually agreed upon contained an important provision adapted from French and German transfer agreements with Pakistan and Brazil, respectively—that transferred technology itself would come under safeguards and that any similar facility constructed by a recipient within a designated period of time would be presumed to involve replicated technology and thus be obliged to be placed under IAEA safeguards.[41]

The process of elaborating these guidelines produced lively (if secret) discussion among the suppliers over preferred strategies to reinforce the nonproliferation regime—notably whether to rely on safeguards and national undertakings, or also to develop and deploy additive restraints (such as technology denial or specially conditioned technology transfers). The arguments did not cause any breaks in the supplier ranks. Rather, they led to a renewed and reinforced consensus on the terms and conditions of doing international nuclear business. Although not fully satisfactory to all of the parties, especially to those like the United States who were concerned about the general effectiveness of a regime dependent on voluntary national undertakings and international verification, the guidelines did reflect an elevated consensus and a sound basis upon which to build a strengthened safeguards system.[42]

Secrecy had surrounded the meetings in London that resulted in the supplier guidelines, and Third World countries were excluded from participation or consultation. Resentment resulted, coupled with charges of cartelism and undercutting of commitments accepted by the advanced nuclear states in Article IV of the NPT. Perceptions and reactions such as these were felt in different ways in important international forums. A Conference on Transfer of Nuclear Technology convened by Third World nations at Persepolis, Iran, just days after President Carter announced new U.S. nuclear policy, revealed a substantial degree of frustration and disappointment with progress in international peaceful nuclear development and with efforts to curtail technology transfer to developing nations.[43] Similar reactions were to be heard at the second NPT review conference (1980) where the unilateral imposition of conditions that went beyond the safeguards required by the NPT was sharply challenged and general complaints were heard about restricted cooperation.[44] In the same year, the United Nations, at the urging of the nonaligned states, voted to sponsor a conference on the Peaceful Uses of Nuclear Energy.[45] Finally, these reactions gave added substance and even a richer agenda to the Group of 77, which has represented Third

World and nonaligned views on North-South issues, technology transfer, economic development, and other matters since the mid-1960s and that made its formal debut in the IAEA at the 1976 general conference held in Brazil. The relevance of this for the IAEA is discussed in chapter 7.

Challenging the Plutonium Presumption

The second element of post-1974 U.S. nonproliferation policy had a rather different twist, and particularly after 1977 became a source of considerable controversy between the United States and its industrial nuclear partners. Unlike the first element, which emphasized voluntary export conditions and limits to transfers of sensitive nuclear technologies, this involved reassessment of a fundamental presumption of civil nuclear power that had guided civil nuclear development from the inception of Atoms for Peace. The decision to take this step came in 1976 when nonproliferation policy and the adequacy of the existing safeguards-based regime became an issue in the presidential campaign.

Impelled by existing congressional pressures and by candidate Carter's emphasis on the proliferation risks associated with anticipated widespread commercialization of plutonium and reprocessing, President Ford, in the waning days of his administration, announced a new nonproliferation policy: henceforth, the United States would not regard reprocessing and plutonium recycling as necessary and inevitable steps in the nuclear fuel cycle and would defer such activities until there was good reason to conclude that the world could effectively overcome any proliferation risks associated with such activities. President Ford declared a moratorium on exports of sensitive technologies and facilities for a minimum of three years, and called upon other suppliers to join the United States in this effort. Plutonium reprocessing issues, including safeguards effectiveness, were to be considered in the framework of a reprocessing evaluation program. Unfortunately, its precise character never was fully worked out before President Ford's term expired, but it emerged in somewhat altered form in the Carter administration in the form of an international nuclear fuel cycle evaluation.[46]

This anti-plutonium thrust conformed closely to the views expressed at the time in an influential private study co-sponsored by the Ford Foundation and the Mitre Corporation, *Nuclear Power, Issues and Choices*,[47] which provided an important intellectual input to President Carter's nuclear policy. Concluding that accessible and economic uranium resources adequate to support the needs of any power reactors that might be built in the twentieth century were available, and that safeguards on sensitive fuel cycle facilities posed enough problems to make their diversion detection ability suspect, the Ford-Mitre report recommended against commercial reprocessing and recycling of plutonium and in favor both of an indefinite deferral of reprocessing and postponement of breeder reactor commercialization. It urged a ban on the export of reprocessing facilities and called for establishment of sufficient uranium enrichment capacity to meet global needs. While these proposals constituted recommendations regarding the future of domestic nuclear energy policy for the United States, they had very profound implications for external policy and international nuclear cooperation.

U.S. Nonproliferation Policy under President Carter

Many of the measures favored by the Ford-Mitre report, subject to some modification, became elements of President Carter's nuclear development and nonproliferation policy. Although his administration did not purport to impose this policy on its cooperating partners (certainly less so than members of Congress would have hoped, as reflected in the 1978 Nuclear Non-Proliferation Act), it made clear that plutonium deferral decisions elsewhere would not only be welcomed but preferred, especially the reassessing of plans for recycling plutonium in thermal reactors, which was the initial fuel cycle activity planned for in commercial use of plutonium.

The greatest proportion of nuclear fuel then being used in reactors in the free world originated in the United States. For reasons discussed earlier, this American predominance in nuclear fuel was undergoing change, and within a few years of the time that President Carter announced his plutonium deferral policy, the proportion of U.S.-origin fuel going into free world reactors would drop significantly. Ironically, U.S. policymakers implicitly relied on fuel supply to help leverage nuclear policies elsewhere, although other means of influence also were available.

In 1977, however, the United States still could control what happened to much of the fuel then in use outside the communist states. In most instances (EURATOM was the exception), U.S. bilateral agreements for cooperation contained a provision requiring prior U.S. consent to the transfer of U.S.-origin fuels for reprocessing or any other purpose. This meant that for all practical purposes other nations were ineluctably caught up in the process and dynamics of the U.S. nonproliferation policy.

Many of these states had made large-scale commitments to nuclear energy and to a fuel cycle development strategy that assumed the eventual use of plutonium as a fuel. Most felt vulnerable because of their generally high dependence on imported fuels and energy, and were still smarting from the effects of the 1974 oil crisis, which had made them all the more determined to reduce their external energy dependence and to use nuclear fuel cycle development (particularly the recovery and use of plutonium) as one important means to that end. None wished to see domestic opposition to nuclear power fed by official doubts in a leading nuclear nation. And some were concerned that their nuclear industries might suffer as a consequence of the convergence of a restrained external market (due to the provisions agreed in the Nuclear Supplier Guidelines) and a shrinking domestic market (occasioned by the impact of policies projected from the United States).

If these considerations alone were not enough to provoke confrontation between the United States and its advanced nuclear state partners, an additional feature of U.S. policy was sure to be the catalyst. Carter administration policy emphasized universality in nuclear policy and sought to avoid discrimination in the implementation of its principles. There were several reasons for this, including the desire to avoid further antagonizing the Third World, which was recognized as important to the future of nonproliferation, and the notion that unless the technological leaders themselves practiced some form of abstention there would be little chance to persuade others of the possible merits of the approach being advocated. The underlying philosophy of this strategy can be found in the Ford-Mitre report, which made the following important argument:

> A United States proposal for international reexamination of the economics of plutonium recycle and breeders will hardly be credible unless the United States is itself prepared to defer its own plutonium recycle and breeder commercialization programs on valid economic and energy supply grounds. Such action will not necessarily convince all countries but will certainly influence their thinking *and will preempt charges of discrimination or of failure to honor NPT commitments.*[48] [emphasis added]

The severity of this confrontation was ameliorated by a number of factors. One was that while advocating universality in principle, in practice, by employing a case-by-case approach, the United States did distinguish between states with specific requests for reprocessing U.S.-origin fuel on the basis of such criteria as their nonproliferation credentials, their adherence to the NPT, and the nature and scope of their existing civil nuclear energy programs. This served to accommodate states that otherwise would have suffered a genuine hardship. Another was the effort of the administration to bring flexibility into the legislation that was then being considered by the Congress so that it could be responsive to legitimate claims of cooperating partners.

A primary example of this was the administration's successful effort in deflecting Congress from adopting an exclusive "timely warning" standard for granting authorization to reprocess U.S.-origin fuel by making this a "foremost" but not exclusive criterion by which to judge requests.[49]

One of the most important ameliorating measures was the initiation of the International Nuclear Fuel Cycle Evaluation (INFCE).[50] INFCE was not intended as a negotiation process (in the sense of resulting in a definitive agreement on the nuclear fuel cycle) but rather as a techno-diplomatic exercise in international technology assessment. It was aimed at addressing the interrelated problems of implementing nuclear power programs while minimizing the risk of further proliferation. On balance, it is generally acknowledged that its main values were to defuse the international nuclear atmosphere; to enable the United States to clarify the nature, scope, and intent of its evolving nuclear policy; and to focus attention on the legitimacy and importance of taking nuclear proliferation into account along with economic and technical considerations in making national nuclear fuel cycle decisions.[51] It provided a forum in which to narrow factual differences and to establish a common analytic base regarding proliferation risk assessment. INFCE also set in motion a process to establish some consensus on a number of significant issues, including alternative management options for spent fuel. And, it contributed to sensitizing the international community to the fact that there are proliferation risks associated with certain fuel cycle decisions as well as to the collective nature of the responsibility for avoiding proliferation by whatever path.[52]

INFCE also was relevant to the IAEA and safeguards. There was wide acknowledgment in the working groups and in the final plenary report of the importance of effective international safeguards for cooperation and development in the peaceful use of nuclear energy. INFCE strongly endorsed IAEA safeguards and their general adequacy for dealing with existing operating plants (primarily light-water power reactors and small-scale reprocessing facilities); it also stressed the value and importance of further development and improvement of safeguards, especially in connection with sensitive nuclear fuel cycle facilities. The central role of safeguards in the nonproliferation regime was seen as justification for continued safeguards research and development to improve methods of materials accountability and enhance means of containment and surveillance.

One cannot leave this brief review of the main themes of U.S. executive branch policy without some comments on the policy's overall impact. All the issues raised were the right ones. The conceptual strategy for dealing with them embraced all of the critical elements, and the effort to engage all interested parties in the assessment process made an international enterprise of American policy. But the approach also suffered from a failure to recognize adequately the depth of interdependence that had developed in the nuclear world since Atoms for Peace (or it recognized the interdependence but misjudged the implications and latitude for action it involved). Actions nominally defined as domestic had immediate, far-reaching, and disruptive effects on many of the United States' traditional nuclear partners and provoked a loss of confidence. They also weakened confidence in the very international institutions (the IAEA and the international safeguards) that they were intended to strengthen. In some respects, it was a case of the best being the enemy of the good.

Reagan Administration Policy and Safeguards

Nonproliferation policy has undergone some change under the Reagan administration, but there is considerably more continuity than sometimes admitted. In part this is because the main differences have been over tactics, not over goals—both administrations share the objective of all postwar administrations, the avoidance of further proliferation. The continuity also exists because implementation of nonproliferation policy, especially in the later years of the Carter administration, was more pragmatic and differentiated than some of the rhetoric of the early years would suggest. Finally, the continuity has been maintained because executive branch policies are constrained by statutory law, in this case the Nuclear Non-Proliferation Act of 1978. Nevertheless, some differences do have a bearing on safeguards and the IAEA.

An important change in nonproliferation policy under President Reagan is the distinction made between states that are and that are not perceived to be reliable vis-à-vis nonproliferation. The Reagan administration has adopted a policy of "continuing to inhibit sensitive transfers of nuclear technology, equipment, and materials, particularly where the dangers of proliferation demand," but not to "inhibit or set back civil reprocessing or breeder development abroad in nations with advanced nuclear power programs where it does not constitute a proliferation risk."[53]

This policy signals a break with President Carter's inclination to deal with nuclear issues on the basis of universal principles applicable to all states regardless of nonproliferation status and credentials. It also diverges from his disposition to question the legitimacy of reprocessing and plutonium use even in existing reprocessing facilities by mature nuclear nations, and with the implicit assumption that effective international safeguards for certain activities (such as reprocessing) are inherently unattainable.

While this safeguards view was not the official position of the Carter administration, it was a view held by a few members of the administration as well as at least one member of the Nuclear Regulatory Commission and several members of Congress. It also was the view of the majority of participants in the Ford-Mitre study, which noted that the nonproliferation system "will inevitably be flawed and unstable if plutonium and highly enriched uranium . . . and the facilities to produce them become increasingly widespread."[54] For others, some reprocessing was viewed as probably inevitable and the disposition accordingly was to seek to make it as safe as possible by strengthening safeguards, limiting the activity to as few sites as possible in politically stable environments, and focusing plutonium use on future breeder reactors rather than on current light water reactors.[55] The inevitability of reprocessing and the desirability of limiting it to stable and mature nuclear states is part of the nonproliferation policy of the Reagan administration as well.

This policy orientation inevitably focuses attention on international safeguards and emphasizes their central role and importance to nonproliferation—a point that was given prominent attention by President Reagan in announcing U.S. policy in July 1981, as well as in urging all supplier states to require full-scope safeguards as a condition of nuclear supply. Indeed, a shift from denial to support, or at least tolerance, of certain kinds of activities involving U.S.-origin fuels is predicated on the existence and application of the most effective safeguards possible. Furthermore, it assumes that not only the United States, but its cooperating partners as well, will provide whatever support is needed to facilitate development and implementation of effective international safeguards in general and of measures to compensate for limitations and weaknesses in current IAEA safeguards in respect to sensitive activities and bulk-handling facilities.

This clearly necessitates adequate funding for the IAEA, agency access to national laboratory research and development and expertise on improved safeguards, and maximum cooperation in implementing IAEA safeguards smoothly, efficiently, and in a timely manner. And as the earlier discussion of legislative activism demonstrated, Congress is disposed to scrutinize closely executive branch policy for nuclear cooperation with a view to ensuring that nonproliferation objectives are fulfilled. Altogether, this adds up to a substantial degree of support and high-level attention being given to IAEA safeguards by the United States.

Critical judgments may be appropriately passed on some aspects of Reagan administration nuclear policy: its seemingly acute emphasis on reestablishing the United States as a reliable supplier; its liberal attitude toward plutonium use; its concept of "bridge-building" to NPT holdouts (which has turned out to be more of a one way street and has not yielded significant results); or its efforts to minimize congressional oversight in the nonproliferation area (for example, with respect to subsequent arrangements incorporated in programmatic arrangements on reprocessing). But the administration does score well on emphasizing robust and effective international safeguards. Administration commitment to the IAEA as a whole is less clear as reflected in the tendency of the United States (more the Congress than the executive branch) to allow other political interests to challenge unequivocal support for the IAEA.

NONPROLIFERATION POLICY AND FOREIGN POLICY

Nonproliferation policy is not synonymous with foreign policy. As we saw earlier in discussing the case of Pakistan, nonproliferation policy objectives may have to compete with other important national goals and interests, and as priorities are sorted out, they may sometimes be supplanted by more urgent concerns. That risk can affect not only specific nonproliferation policies in the United States and abroad but also the IAEA and its safeguards system, as is demonstrated by events after the Israeli attack on an Iraqi nuclear reactor in June 1981.

In September 1982, when the IAEA general conference voted to reject the credentials of the Israeli delegation, the U.S. delegation walked out of the meeting, announcing that the United States would have to reassess its continued participation in the agency. The Senate Appropriations Committee, during the course of the administration's reassessment, voted to delete from an appropriation bill the $14.5 million voluntary contribution intended for the IAEA in 1983, an action that was partially remedied by subsequent legislative action. At the same time, the administration suspended payments of the remainder of its 1982 dues to the IAEA pending the outcome of its reassessment. U.S. support for Israel clearly had clashed head on with the U.S. commitment to the IAEA.[56]

In this instance, the United States concluded that the IAEA was critical to its nonproliferation and national security interests essentially because of the indispensable role played by IAEA safeguards in the global nonproliferation regime. As a result, the United States resumed participation in the agency. This outcome is one instance when nonproliferation concerns were sustained against other values and interests. But it is not a foregone conclusion that this result will always obtain or that the depth of U.S. nonproliferation interest can be held hostage to U.S. acquiescence in actions of others that it regards as fundamentally detrimental to or incompatible with its own national interests. This issue will be revisited in the next chapter.

NOTES

1. On this transitional period, see Bertrand Goldschmidt and Myron B. Kratzer, "Peaceful Nuclear Relations: A Study of the Creation and Erosion of Confidence," in Ian Smart, ed., *World Nuclear Energy: Toward a Bargain of Confidence* (Baltimore, Md., The Johns Hopkins University Press, 1982) pp. 19–48; Onkar Marwah and Ann Schulz, eds., *Nuclear Proliferation and the Near-Nuclear Countries* (Cambridge, Mass., Ballinger, 1985); Lewis A. Dunn and William H. Overholt, "The Next Phase in Nuclear Proliferation Research," *Orbis* (Summer 1976) pp. 497–524; Albert Wohlstetter, "Spreading the Bomb without Quite Breaking the Rules," *Foreign Policy* no. 25 (Winter 1976–1977) pp. 88–96.

2. China, of course, was assisted in its nuclear activities by the Soviet Union, but far less is known about the purpose of that assistance and what Soviet expectations were at the time it was rendered. Thus, India still stands as the only case where clear statements of expectation were made for several years before the actual event, statements that erased any earlier ambiguities in the bilateral agreement between India and Canada.

3. Insofar as the regime consists also of agreements for cooperation, it may be argued that India did in fact violate a regime rule. This argument might be strengthened by the fact that at the time of the Indian explosion the question of the legitimacy of peaceful nuclear explosions under the agreement with Canada was under active discussion, at least making India guilty of unilaterally interpreting a bilateral agreement whose meaning was in dispute.

4. Of course, it would have been possible that a customary rule of international law had emerged with regard to peaceful nuclear explosions. If that were the case, the Indian action would have been more arguably a violation of a regime norm. As suggested here, however, the situation at the time is fraught with too many ambiguities to permit an unambiguous conclusion.

5. For a discussion of Canadian views and policies, see Mark J. Moher, "Nuclear Suppliers and Nonproliferation: A Canadian Perspective," in Rodney Jones and coauthors, eds., *The Nuclear Suppliers and Nonproliferation* (Lexington, Mass., Lexington Books, 1985).

6. See Leonard S. Spector, *Nuclear Proliferation Today* (New York, Vintage Books, 1984) pp. 342–343; Rodney Jones, "Nuclear Supply Policy and South Asia," in Rodney Jones and coauthors, ibid., p. 166.

7. For some, it was a question of misunderstanding just what purpose safeguards were to serve, a misunderstanding that was encouraged by the term itself, which implies some form of hands-on control or the ability to stop an action from occurring. For others, it was less misunderstanding than their having concluded that safeguards were not sufficient to provide adequate protection against diversion and that new and different measures were required. These two views converged in the aftermath of the Indian test to create the situation discussed herein. The problem is reflected in the U.S. concept of "timely warning" as distinguished from "timely detection" that is called for in the IAEA NPT safeguards document INFCIRC/153. Timely detection is detection within a given time specified by the IAEA and is related to material accountancy—it is detection of an event that has occurred. Timely warning, on the other hand, is detection of an event before it has been completed or so quickly afterward that the diverter has no time to convert the material diverted into a nuclear explosive. If one were to use timely warning as the criterion for safeguards adequacy, it is clear that in some cases timeliness could never be achieved. This issue is discussed in chapters 7 and 8, below.

8. French suspension did not occur until most of the plans already had been delivered. Thus, technology transfer had taken place and may have been sufficient to be of real value to Pakistan.

9. The Federal Republic of Germany-Brazil agreement prompted a strong reaction from the United States that resulted in considerable tension between the United States and both of the other countries. For an in-depth review of that situation, consult Norman Gall, "Atoms for Brazil, Dangers for All," *Bulletin of the Atomic Scientists* (June 1976); William Lowrance, "Nuclear Futures for Sale: To Brazil from West Germany," *International Security* (Fall 1976) pp. 147–166; Karl Kaiser, "The Great Nuclear Debate: German-American Disagreements," *Foreign Policy* (Spring 1978) pp. 83–110.

10. Official U.S. statements on this matter can be found in *Department of State Bulletin* vol. 71 no. 1832 (August 5, 1974) pp. 248–254 and vol. 71, no. 1841 (October 7, 1974) pp. 484–486. For a discussion, see Michael A. Guhin, *Nuclear Paradox: Security Risks of the Peaceful Atoms* (Washington, D.C., American Enterprise Institute for Public Policy Research, 1976) p. 31ff.

11. P.L. 93-485, October 26, 1974. Previously, agreements for cooperation were submitted to the congressional Joint Committee on Atomic Energy thirty days before coming into force, but without any provision for disapproval except by legislative enactment.

12. Others included the use of legislative vetoes to check action by the administration that Congress felt to be inconsistent with its policy intent, as well as the growth of congressional committees with interests in nuclear

matters whether from economic, technological, political, security, or environmental perspectives. The legislative veto was struck down by the United States Supreme Court on June 13, 1983, in the case of *U.S. Immigration and Naturalization Service vs. Chada*.

13. Note should be taken of the fact that in 1974, with the passage of the Energy Reorganization Act, the Atomic Energy Commission was divided into a regulatory body, the Nuclear Regulatory Commission (NRC), and a broader energy agency that consolidated all federal energy development programs, the Energy Research and Development Administration (ERDA). What was significant was that the independent NRC was legally free to reach decisions on nuclear exports that were binding on the government. At the same time, other government branches, such as the Arms Control and Disarmament Agency (ACDA), the Department of State, and ERDA, were involved in such decisions. The resulting diffusion of responsibilities and authorities in nuclear matters had a very negative impact on U.S. international nuclear stature. This was in addition to the proliferation of congressional committees that concerned themselves with nuclear matters that resulted from the demise of the Joint Committee on Atomic Energy.

14. See O.B. Falls, Jr., "A Survey of Nuclear Power in Developing Countries," *IAEA Bulletin* vol. 15 no. 5 (August 1973) pp. 27–38. Among the countries included in the survey were Argentina, Bangladesh, Egypt, Korea, Mexico, Pakistan, the Philippines, and Yugoslavia.

15. See in particular, Richard D. Barber Associates, *LDC Nuclear Power Prospects, 1975–1990: Commercial, Economic and Security Implications*, A Report Prepared for the Division of International Security Affairs, U.S. Energy Research and Development Administration (Washington, D.C., ERDA, February 1975).

16. On the character and implications of the decline of U.S. predominance, see William B. Walker and Mans Lonnroth, *Nuclear Power Struggles: Industrial Competition and Proliferation Control* (London and Boston, Allen & Unwin, 1983).

17. This subject is well treated by Michael J. Brenner, *Nuclear Power and Non-Proliferation: The Remaking of U.S. Policy* (Cambridge, England, Cambridge University Press, 1981).

18. This aspect of the episode is very well analyzed and documented in Edward F. Wonder, *Nuclear Fuel and American Foreign Policy*, An Atlantic Council Policy Study (Boulder, Colo., Westview Press, 1977).

19. 10 Code of Federal Regulations Part 810 (Washington, D.C., U.S. Government Printing Office).

20. United States Arms Control and Disarmament Agency, *Documents on Disarmament* (Washington, D.C., U.S. ACDA, 1969) pp. 609–610.

21. See Albert Wohlstetter, *The Spread of Military and Civilian Nuclear Energy: Predictions, Premises and Policies* (Los Angeles, Pan Heuristics, 1976). See

also Wohlstetter, *Swords from Ploughshares* (Chicago, University of Chicago Press, 1979).

22. P.L. 94-329, The International Security and Arms Export Control Act of 1976, June 30, 1976 (Washington, D.C., U.S. Government Printing Office, 1976).

23. P.L. 95-92, The International Security Assistance Act of 1977, August 4, 1977 (Washington, D.C., U.S. Government Printing Office, 1977).

24. How reliable this pledge was deemed to be by the U.S. administration is unclear given the fact that the president did not seek to exercise the waiver under the Symington amendment.

25. On this issue generally, see Leonard S. Spector, *Nuclear Proliferation Today* (New York, Vintage Books, 1984) pp. 70–110.

26. P.L. 97-113, The International Security and Development Cooperation Act of 1981. This type of congressional veto, however, was struck down in 1983 by the Supreme Court in the *Chada* decision referred to in note 12 above.

27. P.L. 95-242, The Nuclear Non-Proliferation Act of 1978, section 2.

28. Ibid., section 2a.

29. Ibid., section 4(a)(6).

30. Atomic Energy Act of 1954, as amended, section 123 (amended and restated by section 401 of the Nuclear Non-Proliferation Act of 1978) deals with Cooperation with other Nations; section 127 (added by section 305 of the Nuclear Non-Proliferation Act of 1978), Criteria Governing United States Nuclear Exports.

31. The Nuclear Non-Proliferation Act of 1978, section 403(b)(1).

32. Ibid., section 104(d).

33. Atomic Energy Act of 1954, as amended, section 129(2)(c) (added by section 307 of the Nuclear Non-Proliferation Act of 1978), Conduct Resulting in Termination of Nuclear Exports.

34. Ibid., section 131(b) (added by section 303(a) of the Nuclear Non-Proliferation Act of 1978), Subsequent Arrangements.

35. On the subject of Japanese reprocessing, U.S. policy and the Tokai-Mura facility, see Lawrence Scheinman and Ryukichi Imai contributions in Michael Blaker, ed., *Oil and the Atom: Issues in U.S.-Japanese Energy Relations* (New York, Columbia University Press, 1980).

36. Programmatic transfers also have been agreed to for Sweden, Norway, and Finland, but unlike the other approvals mentioned, these do not involve the return of separated plutonium. On this issue generally, see "Reprocessing and Plutonium Use," *Department of State Bulletin* (September 1982) p. 52.

37. This includes the proposed Nuclear Non-Proliferation Policy Act (NNPA) of 1982, which in addition to all of the requirements of the NNPA would

have required concurrence of the Nuclear Regulatory Commission in a determination that a proposed subsequent arrangement for reprocessing would not result in a significant increase in proliferation risk; and the Nuclear Explosive Control Act of 1983, cosponsored by Senators Gary Hart and Alan Cranston and Representative Richard Ottinger, which would have prohibited reprocessing U.S.-origin fuel until *Congress* made a finding that effective international safeguards could be applied, and that appropriate sanctions existed to deter any violations.

38. The guidelines eventually were published as INFCIRC/256 as a result of each of the fifteen participants' informing the IAEA by separate letter of their intent to follow the guidelines with respect to their nuclear exports. They include not only a requirement for safeguards, but also a nonexplosive pledge, provision for adequate physical security, and agreement not to retransfer supplied items or their products without approval of the original supplier, and then only under the same conditions as the original supply.

39. INFCIRC/254, paragraph 7, states, "Suppliers should exercise restraint in the transfer of sensitive facilities, technology and weapons-usable materials. If enrichment or reprocessing facilities, equipment or technology are to be transferred, suppliers should encourage recipients to accept, as an alternative to national plants, supplier involvement and/or other appropriate multinational participation in resulting facilities. Suppliers should also promote international [including IAEA] activities concerned with multinational regional fuel cycle centres."

40. France made this announcement in December 1976; the Federal Republic of Germany in June 1977. The Nuclear Suppliers Group is discussed in Charles N. Van Doren, "Nuclear Supply and Non-Proliferation: The IAEA Committee on Assurances of Supply," Report No. 83-202 S (Washington, D.C., Congressional Research Service, October 1983) pp. 60–65.

41. This is a potentially important provision, but a number of people have raised questions about its enforceability.

42. The division of views among the suppliers is discussed in Bertrand Goldschmidt, "A Historical Survey of Non-Proliferation Policies," *International Security* (Summer 1977).

43. The shrillest voices were those of states that had rejected the NPT. States genuinely interested in peaceful nuclear development would not be hurt by these policies and would have access to all of the nuclear technology, materials, and equipment relevant to their level of nuclear development as long as they accepted full-scope safeguards. But these considerations are less important than the fact that this reflected the perception of developing states across the political and nonproliferation spectrum. It is also noteworthy that U.S. industrial representatives participating in the Persepolis conference assisted in drafting the condemnatory resolutions that emerged. ("New Trends and Safeguards," speech delivered by Myron B. Kratzer, vice president, International Energy Associates Limited, to the Institute of Nuclear Materials Management, Albuquerque, N.M., July 1985)

44. For a discussion of the second review conference, see "The Second NPT Review Conference," in SIPRI, *World Armaments and Disarmament: SIPRI Yearbook, 1981* (London, Taylor and Francis, 1981) pp. 297–338.

45. United Nations General Assembly Resolution 35/112. By this resolution, the assembly decided to convene a United Nations Conference for the Promotion of International Cooperation in the Peaceful Uses of Nuclear Energy (PUNE) and to establish a Preparatory Committee of seventy members to that end. The resolution invited the IAEA to contribute to all phases affecting matters within the scope of its responsibilities. PUNE was to have taken place in 1983, but only was convened in 1987.

46. Office of the President, *Public Papers of the Presidents: Gerald R. Ford, 1976* (Washington, D.C., U.S. Government Printing Office, 1977) pp. 2763–2778.

47. Cambridge, Mass., Ballinger, 1977.

48. Ibid., p. 37.

49. Timely warning is discussed in note 7 above. See also, Leonard Weiss, "Nuclear Safeguards: a Congressional Perspective," *Bulletin of the Atomic Scientists* (March 1978) pp. 27–33.

50. In announcing his nuclear policy on April 17, 1977, President Carter stated that pursuant to the objective of continuing discussions with suppliers and consumers over ways to achieve energy objectives while reducing the spread of nuclear explosive capability, the United States wished to explore establishment of an international nuclear fuel cycle evaluation program. The complete statement can be found in *Documents on Disarmament, 1977* (Washington, D.C., United States Arms Control and Disarmament Agency, 1977) pp. 219–220.

51. For a good *post facto* review of U.S. perceptions of INFCE as expressed in congressional hearings on the subject, see Warren H. Donnelly, "Analysis of Hearings on New Directions for Nuclear Energy Research, Development and Demonstration, Post-INFCE," a report prepared for the Subcommittee on Energy Research and Production of the Committee on Science and Technology, U.S. House of Representatives, 97 Cong. 1 sess. May 1981.

52. In the final analysis, it is at least as important that a country that decides to pursue a plutonium use option for peaceful nuclear purposes understand and take all necessary measures to minimize the risks associated with that decision in the first place. The fact that some countries continue to pursue plans involving the use of plutonium does not necessarily support the argument made by some that U.S. policy failed.

53. "Statement on United States Nuclear Non-Proliferation Policy," July 16, 1981, in *Public Papers of the Presidents of the United States, Ronald Reagan, 1981* (Washington, D.C., U.S. Government Printing Office, 1982) p. 631. This was later supplemented by a decision to end the policy of indefinite deferral of reprocessing and plutonium use in the United States and to

endorse those activities. This occurred in October 1981, but industry has not chosen to pursue reprocessing primarily because its economics are unfavorable, but also out of a lingering concern about the stability of U.S. national policy on this issue.

54. Ford-Mitre, *Nuclear Power: Issues and Choices* (see note 47 above) p. 23.

55. See, for example, Joseph S. Nye, "Balancing Nonproliferation and Energy Security," speech to the Uranium Institute, London, July 12, 1978 (unpubd.).

56. The attack on Israel in the IAEA was not an isolated problem, but part of a larger issue of Third World challenges to the United States as well as to Israel and other U.S. allies in the forums of the UN General Assembly and counterpart institutions in other international organizations including the IAEA—challenges to which the United States had determined to respond with toughness as demonstrated by its subsequent withdrawal from UNESCO. These broader issues, while of undoubted importance, go beyond our immediate concern and are not dealt with here.

Chapter 7

PROBLEMS FACING THE IAEA

So long as nonproliferation and sharing the peaceful benefits of nuclear energy remain high on the political and economic agendas of the world, the loss of International Atomic Energy Agency (IAEA) safeguards is unthinkable. However, today the IAEA confronts critical problems that have the potential to undermine agency safeguards and jeopardize their continued utility.

As the IAEA's nonproliferation role has increased, so has its political visibility, which has made it a more attractive arena for political activism. Once able to function in comparative isolation from the dramas of international politics, in recent years the IAEA has become yet another stage for political struggles unrelated to its substantive mandate.

The political campaigns that have been waged against Israel and South Africa in other UN organizations have spilled over into the IAEA, although admittedly helped by such agency-relevant events as the discovery in 1977 of an alleged nuclear test site in South Africa's Kalahari desert and by the Israeli bombing of the Iraqi research reactor in June 1981.[1] Consequently, not only the annual IAEA General Conference, which although largely devoted to routine statements about nuclear programs and experience always has had political overtones, but also the board of governors, which over time has developed rather efficient and businesslike qualities, are being progressively transformed into forums for political debate and confrontation reminiscent of the UN General Assembly and Security Council.

The character of the IAEA has changed over the past fifteen years. As a result of its exceptional authority to implement international safeguards, it has come to occupy a central and indispensable role in the global nonproliferation regime. But this in turn has made the agency an important political prize in the international arena. Groups of states that see the political importance of the agency to other states seek to trade on those interests and to dominate the agency to bend it toward their own preferred goals. In such struggles, the higher purposes of the agency can be lost from sight.[2]

Success inevitably brings its own problems. As the importance of the agency has increased, so have issues of representation, resource development, budgetary allocation, and staffing for IAEA members. In this respect, the IAEA is no different from other international organizations or institutions. However, the IAEA's central role in the nonproliferation regime, which itself is directly linked to questions of international stability and national security, makes all the more significant any changes of behavior within the organization or in how it is perceived by others. As noted above, the IAEA is currently beset by problems that, if unresolved, could fatally undermine the confidence of many states in its ability to carry out its responsibilities. The more important problems are briefly summarized below.

Politicization. One of the most urgent problems is the introduction into agency deliberations and activities of political issues irrelevant to the IAEA's mission, purpose, and objectives. This tends to erode confidence in the organization's ability to carry out effectively its mandated, technically based responsibilities.

Credibility of safeguards. A second, equally important issue is the effectiveness and credibility of IAEA safeguards. This issue involves not only an assessment of how well the agency implements safeguards, but also what purposes and objectives the safeguards should serve.

Tensions between NPT and non-NPT members. Third, IAEA actions suggesting that NPT-related activities are somehow more legitimate than non-NPT-related activities are viewed by non-treaty member states as distortions of the statute and as a threat to the legitimacy of their programs and policies. This has had a corrosive effect on the agency, with non-NPT members resisting actions they regard as attempts to make the agency an arm of the NPT.

The balance between technical assistance and safeguards. The balance between the IAEA's technical assistance program and its safeguards functions, including the question of safeguards financing, is a fourth important issue area. Members from the developing countries and the industrialized nations disagree not so much on whether a balance should exist, but on what sort of balance it should be and how it can best be achieved. This is the issue that most clearly threatens to polarize the agency over the next decade along North-South lines, as has occurred in a number of other international organizations, and already has made some impact on the IAEA itself.

POLITICIZATION

A most serious problem threatening the IAEA is the increasing injection into agency affairs of political issues and controversies unrelated to its charter. For want of a better term, we shall refer to this as "politicization." No international organization is, or realistically can be, entirely free of political debate and controversy. Despite its uncommon record for dealing with subjects on their technical merits, the IAEA is no exception. The agency has had its share of contentious issues and conflicts—East-West confrontations in the early years, and North-South tensions more recently have made their appearance in the Vienna agency. No group of states among its membership can claim to be entirely free of responsibility for such contention. In considering politicization, an important question to ask is whether the political actions involved relate to, or are irrelevant to, the purposes and objectives of the organization. The answer to this inquiry bears on the implications of such activities.

Extraneous and Intrinsic Issues

A useful distinction can be made between extraneous and intrinsic politicization. An extraneous issue involves controversy over a matter unrelated to the mandate of an organization. Disruptive of normal activities, it in no way advances the purposes of the organization. It is politics of a kind that a nonpoliticized organization should immediately rule out of order. As we will discuss later, in its first several years the IAEA reflected the Cold War atmosphere of the general political environment. However, throughout most of its existence, leadership in the IAEA's board of governors and the secretariat have shared a common viewpoint that extraneous issues should be kept out of the agency; the early practice was to refer such issues to the United Nations General Assembly. Occasionally that rule has been breached, but not in a systematic way. More important, extraneous issues generally have not been pressed to the point of disrupting the agency's ability to function or to retain the confidence of its constituents. More often than not, extraneous political interventions have been made more for the record than as part of a dedicated political campaign. However, that is not the situation today.

Intrinsic issues are those that relate directly to an organization's purposes, activities, or structure. These include the allocation of resources among different activities and the distribution of administrative or management posts among different nationalities, or their representation on decision-making bodies of the organization. Intrinsic politicization is an inevitable feature of international organizations. It rarely threatens organizational viability because of the normally strong commitment of the overall membership to the purpose of the organization or to the interests that the organization is effectively serving. If this commitment should weaken, the adverse effects of intrinsic politicization increase. At some point, their corrosive effect could rival the impact of extraneous politicization. There are reasons to be concerned that such trends are operating in the IAEA today.

A Technical or Political Institution?

The IAEA is often referred to as a technical rather than a political institution. In a fundamental sense that is true. It is *not* a general purpose organization, and it clearly was not established to be still another arena for political struggles or confrontation over the great international economic, ideological, or political issues of the day. Rather, it was established for specific and explicit scientific and technical purposes. However, those purposes—to promote the peaceful uses of atomic energy and to implement safeguards—deal with some of the most important technological, political, and security issues of our time. The agency, therefore, cannot escape all politics and political considerations. A Soviet commentator, writing of the IAEA in *Izvestia* three months after the agency began to function, predicted fairly well what the next three decades would reveal: "Whatever the Western delegates may say on the subject, it is clear that the Agency is not an insulated scientific organization divorced from politics and existing in some kind of vacuum."[3] The interdependence of the agency's technical subjects within its security and political environment makes it an attractive target of political opportunism. As a result, there is great temptation for those with a relatively limited interest in the basic purpose of the agency to hold hostage the substantial interest of others in order to secure their own political objectives, whether they be intrinsic or extraneous.

Politicization over South Africa

The treatment of South Africa by its political opponents in the IAEA starkly illustrates the problem of politicization.

Efforts to strike at South Africa because of its apartheid policies date from 1963 when Ghana unsuccessfully floated a proposal to expel the South Africans from the IAEA. At the time, the then smaller and more homogeneous general conference considered apartheid as a political issue that was inappropriate to their technical forum.[4] Thirteen years later, however, the general conference succeeded in ousting South Africa from the board of governors. In adopting a resolution requesting the board to review the designation of South Africa as the "technically most advanced country in Africa"—which heretofore had entitled that country to be automatically seated—the general conference urged the board to take into account "the inappropriateness and unacceptability of the apartheid regime of the Republic of South Africa."[5]

In June 1977, the board voted 19 to 12, with two abstentions, to replace South Africa with Egypt, although Egypt clearly was not the most advanced African nation in terms of nuclear capability. This confirmed that a new form of politics had taken root. In later years, South African credentials were rejected by the general conference,[6] and still later the board prohibited South African participation in the IAEA's Committee on Assurance of Supply, despite the clear relevance of South Africa to that objective.[7] Subsequently, the general conference foreclosed South African participation in any of the committees or working groups of the agency, thus effectively severing all agency cooperation with the government of South Africa. In 1985, the general conference adopted a resolution urging all member states to cease nuclear cooperation of any kind with South Africa,[8] but, as in the past, stopped short of seeking South African expulsion, apparently because the perceived benefits of keeping South African facilities under agency safeguards and extending the scope of those safeguards exceed political antipathy to the South African regime. The United States, while condemning apartheid, has opposed all of the actions described above on the basis of the principle of universality and the practical importance of South Africa to nuclear matters.[9]

Politicization Over Israel

There is political antipathy toward Israel by a number of member states, primarily those of the Middle East with whom Israel is grouped and whose support is necessary for Israel to achieve any elective IAEA office. As a consequence, Israel never has been chosen to serve on the board of governors, despite its advanced standing in nuclear science and technology, and in general has not been able to play as full a role as its technical achievements would have led one to expect. Israel's air strike against Iraq's large research reactor, Osirak, in June 1981—an admittedly extraordinary event—raised the temperature of this festering political hostility greatly and triggered a period of extended political controversy that in one way or another played a predominant role in the life of the agency during the first half of the 1980s.

The Israeli attack was strongly condemned in the United Nations as a violation of the principles of the UN Charter. The Security Council unanimously adopted a resolution calling upon Israel to refrain from further such attacks or threats of attack and to place its nuclear facilities under agency safeguards.[10] The matter also was placed on the agenda of the IAEA board of governors meeting that took place in the immediate aftermath of the incident. There, Iraq unsuccessfully sought expulsion of Israel from the agency.[11]

The board did, however, adopt a condemnatory resolution urging the forthcoming general conference to consider the implications of the attack for the agency and the possible suspension of Israel from the rights and privileges of membership. This was strongly opposed by the United States and several others for reasons that would frequently be invoked in the ensuing four years in response to efforts to suspend Israel or to impose sanctions against her. One was the importance of the principle of universality to the integrity of international organizations in general (as confirmed in UN General Assembly resolutions) and to the IAEA in particular (given its exceptional responsibility for verifying the peaceful uses of nuclear energy). The other was the absence of appropriate statutory grounds for taking proposed measures against Israel. Article XIX of the Statute of the IAEA, which deals with suspension of rights and privileges of member states, can be invoked only if there has been persistent violation of the statute or of an agreement entered into pursuant to the statute, for example, a safeguards agreement. Neither the attack on the Iraqi facility nor Israel's failure to accept safeguards on all of its nuclear activities constituted a violation of a statutory provision or of an agreement entered into pursuant to the statute. The statute contains no provision regarding the use of force against nuclear facilities whether or not under safeguards; nor does membership require acceptance of safeguards. Hence, Article XIX did not provide any legal basis for suspension or for other punitive measures against Israel.[12]

Iraqi-led efforts to achieve suspension of Israel at the 1981 and 1982 general conferences failed, but lesser measures did receive support, confirming that a majority of the membership believed that the incident was an appropriate one for agency consideration. In 1981, the general conference adopted by a vote of 51 to 8 (which included the United States) with 27 abstentions (mainly European states) a resolution condemning the Israeli attack. The resolution urged that at its next session the conference consider suspending Israel from its rights and privileges of membership if by then it had not withdrawn its threat to attack Iraqi or other nuclear facilities, and placed its own nuclear activities under IAEA safeguards. It also called for immediate suspension of technical assistance to Israel.[13] Technical assistance was suspended. The action, although largely symbolic given the modest level of support involved—about $40,000—was nevertheless important insofar as it affected the principle of full participation of member states. That would become an issue at the 1985 general conference in the context of an Iraqi resolution seeking to invoke sanctions as discussed below.

The Israeli issue came to a head at the following general conference in September 1982. Renewed efforts to suspend Israel were once again unsuccessful. A resolution sponsored by the Moslem states and Cuba received a majority of the votes cast but was defeated because a vote to suspend required a "yes" vote by two-thirds of those voting (excluding abstentions), and the draft resolution in question fell short of that mark.[14] The supporting statements of the sponsors and the language of the draft resolution itself reflected the degree to which political considerations unrelated to the agency or the Iraqi incident had entered the debate. As expressed in preambular paragraph (f) of the proposed resolution, the sponsors were "deeply concerned about Israel's escalation of aggression against other countries of the region and occupation and annexation of territories, particularly the occupation of Lebanon and the genocide perpetrated against the Palestinian people. . . ." They called for suspension of Israel's privileges and rights of membership.[15] Israeli relations with South Africa also were invoked as a reason to take the recommended action against Israel.

Following the defeat of the effort to suspend Israel, Iraq moved to seek rejection of the credentials of the Israeli delegation. Once again, political considerations irrelevant to the mandate of the agency or to the question at hand were introduced. One claim was that the credentials were issued by a government that purported to represent people over whom it exercised jurisdiction by virtue of illegal annexation (referring to Jerusalem) and therefore were deficient. With the assistance of an erroneous legal ruling by the secretariat's chief legal officer, which permitted the casting of a questionable vote, they succeeded.[16] When first put to a vote, this resolution did not succeed. The vote was 40 "yes," 40 "no," plus a number of abstentions. After the vote was announced, a ruling permitted a member state, Madagascar, to reopen the voting and cast a late vote, which altered the outcome. Cases involving credentials only require a simple majority to pass. Coming as it did in the final hours of the conference, the practical effect of the vote was to deprive Israel of participation in only a few remaining activities; but while Israel remained a member of the IAEA, the symbolic impact was not that far from what would have been achieved by a vote of suspension, and it was ominous for the future.

This action against Israel precipitated U.S. withdrawal from the conference that was joined by some others. Altogether, fifteen states withdrew from the conference in protest against the incident, among them France, Belgium, Japan, Italy, and the United Kingdom. At the same time, the United States announced that it would halt general participation in agency activities, including payments of its contributions, and undertake a major reassessment of its support for and participation in the IAEA. This reassessment was completed four months later with the conclusion, endorsed by the president, that the IAEA serves critical U.S. security and nonproliferation interests. Participation was resumed in February 1983 following certification by the director general, on the authority of the board of governors, that Israel remain a fully participating member of the agency.[17]

The intensity of the U.S. reaction had a somewhat sobering effect on the membership of the agency, but it did not put an end to politicization or to bickering over the Israeli issue. Iraq and its supporters shifted emphasis from seeking suspension of Israel to threatening to impose sanctions if Israel failed to take certain measures. In 1983, the general conference voted a resolution (49–24 with 17 abstentions) that called upon Israel to "withdraw forthwith its threat to attack and destroy nuclear facilities in Iraq and in other countries," and further "decided" certain sanctions (withholding research contracts, not purchasing equipment and material, and not holding scientific seminars in Israel) if Israel did not comply and remove its threat by the time of the 1984 general conference.[18] The United States voted against this resolution (RES/409).

In May 1984, Israeli Prime Minister Yitzhak Shamir, singling out the IAEA, publically stated that "Israel supports those international arrangements which would ensure the status and inviolability of nuclear facilities dedicated to peaceful purposes." Shortly afterward the head of the Israeli AEC wrote to Director General Blix that "Israel holds that nuclear facilities dedicated to peaceful purposes be inviolable from military attack," and that Israel "has no policy of attacking nuclear facilities dedicated for peaceful purposes anywhere."[19]

The United States and several others viewed these statements as sufficiently responsive to RES/409 to be able to put the issue to rest, but a number of others, including some who considered the statements to constitute a positive step, disagreed, noting the ambiguous nature of the Israeli statements regarding future behavior, the lack of specific reference to Iraq, and the absence of any mention of the role of IAEA safeguards in providing assurances about peaceful uses. The 1984 general conference voted a somewhat milder but still condemnatory resolution (RES/425) demanding that "Israel undertake forthwith not to carry out any further attacks on nuclear facilities in Iraq or on similar facilities in other countries, devoted to peaceful purposes," and mandating the director general "to seek personally from the Government of Israel" the undertakings in question and to report back to the 1985 general conference.[20] This resolution effectively deferred implementation of any punitive measures against Israel and left intact Israel's privileges and rights of membership while still not bringing the issue to term.

The director general's efforts resulted in a letter from the government of Israel in September 1985 reiterating the stated policy of Israel that peaceful nuclear facilities be inviolable from military attack and Israel's respect for how the IAEA fulfilled its safeguards mission. It also included the following key statements:

> Israel holds that all states must refrain from attacking or threatening to attack nuclear facilities devoted to peaceful purposes, and that the safeguards systems operated by IAEA brings evidence of the peaceful operation of a facility.
>
> It is within this context that Israel reconfirms that under its stated policy it will not attack or threaten to attack any nuclear facilities devoted to peaceful purposes either in the Middle East or anywhere else.
>
> Israel will support any subsequent action in competent fora to work out binding agreements protecting nuclear installations devoted to peaceful purposes from attack and threat of attack.[21]

This important statement went further than previous Israeli statements in two respects: first, it was more direct and precise; and second, it emphasized the role of agency safeguards in evaluating the peaceful nature of nuclear activities, thus taking a major step toward meeting earlier complaints that Israel was putting itself above, and discounting, international safeguards.

Iraq nevertheless remained dissatisfied and introduced a resolution once again demanding that Israel withdraw its threat to repeat its military attack against Iraqi nuclear installations and invoking the punitive measures called for in paragraph 3 of RES/409.[22] The conference president, supported by the legal adviser to the conference (both from African states), ruled that since the resolution in question would have the effect of depriving a member state of some of its rights of membership, its adoption required a two-thirds majority. Iraq's challenge to this ruling was defeated and the resolution, when put to a vote, failed to achieve the required two-thirds. (The vote was 41 to 30, with 19 abstaining.) Instead, the conference adopted (by a vote of 30 to 21, with 36 abstaining) a Nordic-sponsored resolution that, taking note of the Israeli letter and subsequent confirmatory statements before the general conference, "considered" them to contain undertakings meeting the objectives of RES/425, "noting" that thereby Israel has "committed itself not to attack peaceful nuclear facilities in Iraq, elsewhere in the Middle East, or anywhere else."[23] Ironically, the United States did not vote in favor of this resolution because it contained other provisions involving questions the United States regarded as beyond the purview of the IAEA.

With the defeat of the Iraqi draft resolution and adoption of the Nordic alternative, a majority of the members of the agency appeared to be saying enough is enough and that the corrosive five-year debate should go no further. Important national interests served by the agency were being threatened, and steps had to be taken to avoid further deterioration. Thus, the Israeli issue seems at last to have been put to rest in the IAEA, at least insofar as it was based on the Iraqi incident. This does not, however, signal the end of politics or of the continued risk of politicization in general in the agency any more than it suggests that Israel will now be able to overcome the more subtle limitations imposed on her by political considerations. There are, as we shall see, other issues from which political tension and controversy may arise. Nevertheless, one of the most distressing periods in the life of the agency had passed.[24]

In concluding this discussion of the Israeli issue, a final point is in order. The outcome could be viewed as confirmation that the U.S. withdrawal from the agency in 1982 (and its subsequent threats to do so again if Israel were denied the rights and privileges of membership or otherwise dealt with punitively) was a success and a victory against politicization. However, to reach such a judgment would be to indulge in some measure of self-delusion. First, there was a good deal of resentment toward the United States among its friends and allies for the continued threat of withdrawal. In the view of many,

U.S. behavior in this regard contributed to the very politicization the U.S. professed to want to end. The threat to withdraw was seen as unnecessarily painting the United States into a corner, as for example expressed in a letter from the Department of State to Director General Blix in 1984:

> Should the General Conference adopt a resolution that directs the Board . . . to withhold research contracts to Israel, discontinue the purchase of equipment and materials from Israel or refrain from holding seminars, scientific and technical meetings in Israel, the United States Delegation will leave the Conference and announce the suspension of U.S. participation in and support for the Agency. This is a firm and non-negotiable policy.[25]

Efforts by others to avoid passage of resolutions that entailed any punitive measures were mounted less because of convictions that any such action constituted an unacceptable degree of politicization than because of recognition that the United States might just walk out a second time (having already demonstrated in 1982 that it would do so) and that if it did it might be much more difficult, if not impossible, to climb back on board again. That would mean the demise of the agency, which was not in anyone's interest. In this view, rather than eliminating politicization, U.S. behavior might demonstrate that going to the mat is the best way to achieve one's goals, and that could intensify rather than diminish the risk of politicization.

Second, the Damoclean threat of withdrawal raised questions about U.S. priorities. If the United States were prepared to withdraw if any punitive measures were levied against its Israeli ally, then what was one to think about all the U.S. protestations about the importance of maintaining a strong and vigorous nonproliferation regime and an effective international safeguards system? In the Israeli episode, U.S. behavior showed that the importance it attached to nonproliferation policy was distinctly in second place when it came to supporting Israel. The well-known fact that this policy was in no small measure driven by a Congress extremely sensitive to questions relating to Israel, and that the pro-Israeli congressional constituency was larger and more effectively organized than that for nonproliferation, explained but did not resolve the issue for America's allies. Clearly, if the United States could set certain national interests above the IAEA, other members could do so as well.

Finally, many of those who agreed with the United States that it was not for the IAEA to apply sanctions or to suspend Israel in response to the attack on the Iraqi nuclear facility nevertheless disagreed with the view that the incident did not constitute an attack on the agency or on its safeguards system, that it had not done damage to that regime, or that it was inappropriate for the agency to discuss the issue of armed attacks on nuclear facilities.[26] Quite the contrary, many nations that supported U.S. efforts to foreclose punitive action against Israel nevertheless saw the incident as a "grave challenge to the Agency's safeguards system"[27]; causing "serious harm to the Agency's safeguards system"[28]; and as an incident that "could seriously undermine respect for the safeguards system."[29] These and other states share interest in building legal and political barriers against attacks on safeguarded nuclear facilities and in pursuing such issues in the framework of the IAEA as well as elsewhere. At the very least, they are not prepared to foreclose discussion of the issue in the agency.

Other Controversies

In the short term, the controversy over Israel was the gravest and most immediately divisive force within the IAEA. But it was not the first, as we already have seen, nor will it be the last. In the longer term, there is the emergence within the IAEA's governing bodies of well-organized political groupings whose strivings can increase the level of political divisiveness as well as displace the tradition of seeking accommodation through consensus-building with confrontational politics, including demands for formal voting on all issues. Also, as world economic difficulties continue, an increasing North-South polarization in agency proceedings can be expected. This source of politicization opens the door to extraneous politics. It is reflected in confrontation politics on such internal issues as increasing Third World membership on the board, providing larger and more reliable funding for technical assistance, and pressing for greater representation of Third World countries on the secretariat staff, particularly at managerial levels.

Bloc Behavior and the Group of 77

Many groups of states are involved in some way with the IAEA. These include the Geneva Group, the Latin American Group, the West European and Other Group (WEOG), and the Group of 77 (G-77). Often their memberships overlap, and most can be found in other UN organizations as well. The Geneva Group, consisting of the major non-communist contributors to international organizations, provides a means to achieve consensus on budget policy. The WEOG and the Latin American Group are more general-purpose entities, the WEOG functioning with varying levels of intensity in UN family organizations.

Far more important for the future of the IAEA is the Group of 77. The G-77 was formed in 1964 at the UN Conference on Trade and Development. Its purpose is to formulate and aggregate the views of the less-developed countries on matters of economic development and to press for their achievement in the UN system. The G-77 states seek to project their international political agenda, which since the late 1970s has included establishment of the New International Economic Order, in the organizations to which they belong.

The effect of G-77 on the IAEA has been (and continues to be) twofold. First, wherever this caucus has become active it has tended to result in group voting. The G-77 threatens to disrupt the tradition of the agency's functioning largely by consensus and without formal voting, except as provided for in the statute or on rare occasions when demanded by one or more states. It may well alter a style of cooperation that has worked well over the years for the IAEA, one that so far has successfully avoided this kind of group politics. (Of course, G-77 spokesmen may retort that there is nothing sacred about the consensus process if all it does is perpetuate political, economic, or technological arrangements that one seeks to alter.) Second, while the G-77 agenda contains legitimate concerns that merit serious attention by the full IAEA membership (the intrinsic issues), its activity also has been focused on extraneous issues that generate controversies that can threaten the integrity of the IAEA. Unfortunately, IAEA membership includes countries that do not place a priority on the agency's nonproliferation objectives, at least as they have been interpreted by the majority in recent years.

The G-77 can have a constructive voice in the agency if its leadership is drawn from those moderate states with serious nuclear interests and a strong commitment to the principle of nonproliferation and effective international safeguards. However, its leadership currently rests with a few countries—mainly those not party to the NPT—that have opposed the trend toward increased attention to safeguards. These nations are supported by others whose priorities lie less in nuclear advancement per se than in furthering the general objective of establishing a new international economic order, including access to a spectrum of advanced technologies and a redistribution of global wealth.

This is reinforced by the proximity of the IAEA to the UN Industrial Development Organization (UNIDO), which shares the Vienna International Center complex with the IAEA, and whose secretariat views its mission as one of advancing the cause of a new economic order. Often, the same individual will represent his country in both organizations, with priority of interest and balance of expertise favoring UNIDO. There is a resulting tendency to carry over into the IAEA some of the practices and attitudes characteristic of UNIDO: a more confrontational approach to issues; greater attention to symbolic than to substantive aspects of problems; and a greater focus on altering the structure of representation on the board of governors and in the secretariat staff. Establishing majorities sometimes appears to be a more important goal than advancing the technical purposes of the agency.

Efforts by the Group of 77 to Increase Influence on the Board

One of the most striking and important features of the IAEA is its board of governors, which is vested with more executive authority than most comparable UN institutions. As discussed in chapter 3, the board's composition is set by a formula that tries to balance the interests of advanced and developing countries and of East and West. Thirteen of its present thirty-five seats are assigned, on the basis of advanced nuclear status, to IAEA member states by outgoing board each year. These are tantamount to permanent seats for those who hold them. The remaining twenty-two are elected by the general conference for two-year terms in such a way as to ensure a prescribed regional distribution.[30] As a result of pressures from developing countries, the statute, amended in 1963 to increase board membership from twenty-three to twenty-five, was amended again in 1973 to increase the board to thirty-four in response to a convergence of pressures from developing countries as well as a few advanced states.[31] It was increased to thirty-five in 1984 specifically to accommodate the People's Republic of China, which had just taken its seat in the agency.

Pressures have been mounting since 1977 to expand the board even more generally to reflect the IAEA's regional membership, especially in Africa and the Middle East, the two areas from which many former colonial states have emerged with powerful incentives to achieve advanced technological standing to match their new independent status. This has been resisted by the United States and other developed states, including the Soviet Union, on the grounds that the board, as structured at present, is large enough to be representative of the cross-section of its membership while being small enough to operate efficiently, a quality essential to the nature of its tasks and its consensus mode of operation. A further increase in size is seen as potentially jeopardizing the ability of the industrialized countries to prevent decisions inimical to them on matters that require a two-thirds majority (the so-called blocking third) such as approval of the budget and selection of the director general. In a highly politicized atmosphere, the possibility that major agency members would lose their veto takes on even greater significance. Some developing states, particularly from Latin America, also have opposed enlargement out of concern that any alteration could have the politically undesirable effect of diminishing their proportionate representation on the board.

However, the issue has been routinely pushed by the G-77 in a series of general conference resolutions and in consultations within the framework of the board. As in the past, the permanent board members, though unable to quell the demand for change, have been able to channel it along lines that avoid political rupture. In no small measure this has been due to the fact that any proposal holds the potential for disadvantaging one or another group, thus undermining the emergence of the consensus necessary for change. The inability to resolve the issue of elected seats has led some states (Spain, Belgium, Italy, and Sweden) to support the notion of revising the relevant article (VI) of the statute as a whole.[32] In part, this reflects the interest of some of these more advanced states to avoid a situation in which they would end up underrepresented because of some future agreement on enlargement; in part, a desire to assure themselves a more permanent status on the board. This approach has not thus far significantly advanced the dialogue on representation, and the matter remains on the agenda of the board for further consideration. It bears notice that despite the keen interest of a number of states in achieving expansion, and frustration over lack of progress, the 1985 general conference has left the issue to the board for the moment.[33]

Attempts to Increase the G-77's Presence in Key Staff Positions

National claims on appointments are common to all international organizations, and the IAEA is no exception. The issue got caught up in the process of selecting a new director general in 1981. After months of intense lobbying and negotiation, the candidate preferred by the developing countries lost out to the current director general, Hans Blix. In accepting this choice, the general conference, at the initiative of the G-77, passed a resolution enjoining the new director general to "increase substantially the number of staff members drawn from developing areas . . . particularly at the senior and policy-making levels."[34]

The position of the G-77 has become a confrontational issue in the agency, not because there is any serious argument on the merit of the developing world's case that they are underrepresented on the agency's staff, but because of the concerns expressed by developed countries that the quality of staff, not proportional representation, should be the key consideration in all appointments. It is argued that in light of the complex technical character of its activities, IAEA performance would suffer if its appointments were governed by political principles rather than by the best interests of the agency as a whole.

Since 1981, the director general has been responsive to the resolution, occasionally drawing the ire of some advanced countries whose own candidates have been passed over in the process.[35] Both in terms of total professional staff and of senior and policy-making staff, the proportion of developing country representation has increased. Whereas in 1981 slightly more than 15 percent of total staff were from developing countries, this had increased to over 22 percent by 1985. In absolute terms, there were 128 developing country nationals on staff in 1985 (out of 574 total professional staff) compared with 75 (out of 481) in 1981. That is to say, more than one-half of the newly recruited professional staff in this period were drawn from Third World nations.[36]

As for senior level personnel (director and deputy director general level), nine of twenty new appointments in the period in question came from developing countries and the total percentage of senior staff from those countries increased from just over one-quarter to just under one-third. While even these advances have left a number of developing countries less than satisfied, the general sense is that the agency is on the right track, that the trends are positive, but that further efforts are needed and the secretariat and director general are on notice that Third World nations will be watching expectantly to see how recruitment of professional staff is managed over the coming years, with the risk ever present that a perceived unrequited imbalance will generate renewed conflict.[37]

Prospects for Success of the Group of 77

For the moment, the G-77 epitomizes the North-South split, supplanting the East-West division that troubled the IAEA in its early years. For awhile, a superpower condominium existed in which the interest in safeguards predominated. Safeguards remain central, but the time of condominium has passed. The agenda of the current period, which is still somewhat inchoate, promises to be more characteristic of the general North-South dialogue. It is likely to be reflected in the IAEA not only in program emphasis, but in the character and composition of the principal decision-making bodies of the agency. As matters now stand, however, no single group on the board is able to muster a controlling majority alone. However, with support from the Soviet bloc, members of the G-77 on the board would be in a position to control the outcome of a large number of issues brought before it for decision, especially if consensus decision making were to give way to formal voting. Significantly, a major factor inhibiting the G-77's effectiveness in mobilizing such a majority is Soviet agreement with the United States and other Western states on many controversial issues. This largely is a reflection of the Soviet Union's interest in effective international safeguards.

It is conceivable that an enlarged Third World group that maintained its solidarity might achieve its objectives alone. Whether the G-77 can bring about fundamental changes is still unclear, however. The group is not as large in the IAEA as it is elsewhere in the UN system, although more than a dozen potential members are waiting in the wings, held back in some cases only by their inability to pay membership costs.

Furthermore, the G-77 is less homogeneous or cohesive in Vienna because of a number of internal cleavages. There is the difference in nuclear standing among its members; some G-77 states are advanced, others are moderately so, and some not at all. Some wish to use nuclear power, others want to emphasize non-power uses such as radioisotopes and radiation techniques, and still others have no evident nuclear interest. Finally, some G-77 countries are parties to the nonproliferation treaty and anxious to reap benefits from having joined the treaty (for example, Mexico, the Philippines, and Egypt), while others are nonsignatories to the NPT and do not share these views (for instance, India, Pakistan, and Argentina). Thus, for the moment, the G-77 lacks the cohesiveness necessary to exercise a dominant political force within the agency. Nevertheless, all G-77 members see themselves to be on the other side of the fence from advanced nuclear states when it comes to issues of technical cooperation and technology transfer. And, because of this, they have the sense of solidarity often felt by the have-nots.

Possible Consequences of G-77's Domination

What are the long-term implications of a large, coherent G-77 presence in the IAEA? Perhaps it is instructive to consider what could happen if the G-77 dominated the board. For example, the G-77 could then—in competition for resources—emphasize promotional and technical assistance activities, possibly at the expense of safeguards.[38] Many of the G-77 countries feel less urgency about international safeguards since they see no serious regional nuclear threat to themselves. Some states, in Africa for example, are inconsistent. On the one hand, they are inclined to downplay the importance of safeguards when it is a question of scarce resources. They perceive support for increased safeguards (for which they won't have to pay in any event because of safeguards financing formulas that have been established to shield them from the effects of increased safeguards costs) as threatening the resources that may be available through voluntary contributions for technical cooperation. On the other hand, they understand that their security is better preserved if South African nuclear activities are under safeguards than if they are not. Others, such as India and Pakistan, face regional threats, and seem to prefer ambiguity and keeping open their nuclear weapons options rather than accepting comprehensive safeguards and formal NPT-type undertakings. Some of the G-77 might be tempted to seek to apply agency safeguards so as to harass the unpopular states, thus turning safeguards away from their purpose of confidence building and threatening their continued acceptability. At least for these reasons, it is urgent to make every possible effort to reduce controversy in the IAEA and to enlist the active involvement of moderate members of the G-77 who are seriously interested in peaceful uses of nuclear energy, to reinforce the constructive, cooperative, and consensus-based qualities of the Vienna agency.

Politicization from the West

It would be misleading to leave the impression that politicization is the result of Third World activism alone. That is not correct either historically or in the contemporary situation. As noted earlier, while the IAEA was largely free of extraneous politics over much of its life, this was not entirely so. Cold War issues—such as the admission of the People's Republic of China, Hungary, North Korea, and East Germany—were systematically raised by the Soviet Union and just as systematically rejected by the United States.[39] Eastern bloc countries opposed technical assistance to South Korea, South Vietnam, and Taiwan. On the grounds of alleged Israeli aggression in the Middle East war, Arab states in 1967, with Soviet support, tried to block the transfer of equipment awarded to Israel on the basis of technical merit. On its part, the United States, following a congressional mandate applicable to all international organizations, each year reduces its voluntary contribution to the IAEA by its proportionate share of whatever technical assistance is provided to Cuba and, until recently, Vietnam.[40]

Complaints about politicization were not heard earlier in the West because the Western nations had the upper hand and dominated the general conference and board of governors. The view from Moscow was undoubtedly different. U.S. predominance in the general and more specific nuclear energy arena has since declined, and the influence of other groups of states has risen along with their numbers.

As U.S. influence has diminished, its insistence on principles such as universality—to which it adhered less vigorously when it was in a more favorable position—appears to have strengthened. This is evident in the Israeli policy (discussed earlier) where we noted that some close allies of the United States saw the insistence of universalism more as a political argument to support national policy on Israel than as a true commitment to universalism in international organizations. Some, for example, have watched with interest to see whether the United States would respond to rejection in other cases, such as South Africa, with equal vigor as it did for Israel. These criticisms, even if correct in their assessment of U.S. motivations and selectivity, overlook the importance of the underlying principle that U.S. actions defend—that any effort to restrict the rights of individual members diminishes the entire organization.

Another aspect of the U.S. contribution to politicization is the nostalgia for the "good old days," when the agency was allegedly virtually free of political considerations and devoted itself exclusively to technical and scientific matters. As we have seen, that perception does not accurately reflect the true situation. The problem today is that too much insistence on returning to an ideal situation that never quite existed can strain the structure and membership of the agency, thus generating more problems than it solves. In any event, the fact is that the political climate of the 1980s differs from that of the 1960s and must be dealt with on its own terms rather than on those that were more applicable to a bygone time and political situation.

THE CREDIBILITY OF SAFEGUARDS

The credibility of safeguards involves both conceptual and operational questions, several of which were introduced in the two earlier chapters describing the IAEA's safeguards system. In this section we will focus attention on the most important problems affecting the credibility of international safeguards.

Overexpectation for Safeguards

Beginning in the mid-1970s, India's nuclear test and other events (discussed earlier) generated concern that the NPT was no longer sufficient, and that the overall regime could not assure that further proliferation would not occur. Various public interest group spokesmen and academics argued that the imperfections and fundamental limitations of safeguards were sufficient reasons to forgo the commercial production and use of plutonium as a nuclear fuel, if not to abandon nuclear energy entirely. Their argument in large measure turned on a definition of safeguards that went beyond that of the IAEA statute and reflected differing views of their purpose and objective. Whether this overexpectation was innocent or cynical, it has had and continues to have a powerful effect.

A close observer of international nuclear affairs noted in 1976 that:

> On the whole the greatest difficulty faced by the IAEA is overexpectation. The Agency is not and cannot be a magical way to relieve the world of the problems of proliferation. It is not superhuman, nor should it be. The Agency is a finite, human institution that exists with the forbearance and support of its member states. It has the strengths and weaknesses of any international organization. Overexpectation . . . can lead to unjustified disappointments, malaise and frustration and may discourage practical actions by . . . member states to make the Agency more effective in dealing with proliferation.[41]

Overexpectation remains a central problem for IAEA safeguards. Perhaps this should not be surprising in view of the emphasis placed on safeguards in the international nuclear nonproliferation regime. With so central a role, it was inevitable that safeguards would be closely scrutinized for effectiveness and credibility. But while one might reasonably have expected different levels of tolerance of achievement of commonly understood objectives to emerge, there was less reason to anticipate that differences would emerge as to the purpose and objective of safeguards per se.

Prevention or Verification?

The fact is, however, that a major difficulty in coming to grips with the question of safeguards credibility is the absence of consensus on their purpose, which can be stated broadly as whether prevention or verification is at issue. Some critics of policies for controlled nuclear cooperation and of the nonproliferation regime tend to equate safeguards with the regime and to blame them for weaknesses of the regime. Those following this school of thought tend to judge the efficacy of safeguards by their capacity to prevent proliferation by preventing the diversion of nuclear material. Under this approach, it is reasoned that if international safeguards cannot accomplish this, they should be considered of limited value, if not worthless.[42]

This preventive concept of safeguards is close to the ideas of the Acheson-Lilienthal Report and the Baruch Plan, which aimed at preventing *any* nuclear weapons anywhere; they urged the more radical approach of international ownership and control to international inspection of nationally owned facilities. Such a conception argues for international control over the production and accumulation of nuclear material as well as safeguards to deter diversion. For partisans of this approach, the agency's safeguards should extend to physical protection of nuclear materials and facilities. But in reality this is not possible: in a world of sovereign states, physical protection is the responsibility of nations through the exercise of their police powers, which (with the exception of the veto-controlled United Nations Security Council) is an authority denied to intergovernmental international organizations. This preventive concept also implies that the reach of safeguards should extend to diversion or theft by individuals or by criminal or terrorist groups. But this, too, remains a national responsibility, not an international one. Some who share this view go even further, arguing that if nonproliferation means preventing the acquisition by a non-nuclear weapons state of both nuclear weapons and weapons-usable material (notably plutonium or highly enriched uranium), then the effectiveness of safeguards must be judged against their capacity to prevent such acquisition. From this point of view, the agency should not only sound the alarm in case of diversion but should itself also put out the fire even though it has no fire-fighting equipment.[43]

Most conceive of safeguards as only one element, albeit a crucial one, of the international nonproliferation regime and see their purpose as verification and confidence building. This is the view favored by the agency. Its current director general has frequently emphasized that IAEA safeguards are a unique verification system in which governments invite on-site inspection to demonstrate that no nuclear material is being diverted, and that material under safeguards can be accounted for.[44]

Expanding on this theme, IAEA safeguards may be seen as involving three interrelated purposes—verification, deterrence, and detection. In these terms, verification enables safeguarded countries to provide objective assurance to others that they are adhering to their stated obligations and undertakings. Conversely, safeguards can give other states assurance that their assessment of the nature of nuclear activity in the neighboring safeguarded state is correct. This of course presumes that the agency's safeguards activities are perceived as adequate to make the requisite finding of no diversion. Perceptions of adequacy in turn are based on availability of information about the safeguards inspections, findings, and results.[45]

A second, and related objective in this perspective is to deter governments from violating their agreements that they will not divert safeguarded material to nuclear weapons. The basis for this deterrence is the risk of detection and the political and related costs that detection might entail. Not all proponents of safeguards-as-verification endorse the notion of safeguards-as-deterrence. Some feel that since states voluntarily assume safeguards commitments, it is illogical to suggest that safeguards must somehow be required to deter them from doing what they already have agreed not to do. Others feel that emphasis on deterrence does more harm than good, since most states are fulfilling their commitments and resent any implication that they are not. This is more of a semantic than a substantive issue, however, because the safeguards activities needed to provide verification assurance are virtually the same as those involved in establishing a risk of detection sufficient to deter a diversion. In both cases, the agency must be (and more importantly *be perceived to be*) able on a timely basis to detect a diversion should it occur.[46]

A third objective—and this is the most tangible, but also perhaps the most misunderstood of the three—is to detect diversions of safeguarded nuclear materials if and when they should occur. It is the most tangible criterion in the sense that outsiders most readily associate detection with an identifiable event or incident. But it is misleading because safeguards are based primarily on material accounting, an auditing procedure, which means that "detection" of an event, which may or may not mean a diversion, is a statement about an event that already has occurred.[47] Many people, however, associate detection with a burglar alarm that sounds at the very instant that a theft takes place. Indeed, that is the meaning of the "timely warning" concept incorporated in the United States Nuclear Non-Proliferation Act of 1978: there, "timely" essentially means warning in time for diplomatic efforts to prevent production of a nuclear explosive.[48] As the proponents of safeguards-as-prevention in particular are quick to note, certain materials, such as separated plutonium and highly enriched uranium, if diverted, in principle could be used so quickly to make nuclear explosives that no timely warning would be possible;

and therefore they advocate foreclosing any nuclear activities involving those weapons-usable materials.

The problem of defining the purpose of safeguards is not limited to prevention versus verification; or whether priority should be given to verification, deterrence, or detection; or whether the purpose is assurance rather than deterrence. It also extends to less abstract arguments, reflected in the fact that safeguards are means to an end, not ends in themselves, and that those ends are defined by the underlying treaty or agreement upon which a safeguards agreement with the agency is predicated. In the case of safeguards agreements negotiated pursuant to the NPT, the end is nonproliferation and the obligation is to accept safeguards on all source or special fissionable material in all peaceful nuclear activities for the purpose of verifying that such material is not diverted to nuclear weapons or other nuclear explosive devices.[49]

In the case of safeguards agreements negotiated outside the NPT, or the Treaty of Tlatelolco, the ends invariably are more narrowly defined (that is, they do not entail a general nonproliferation commitment) and the underlying obligation normally relates only to materials or equipment that are provided by a supplier to a recipient, not to all nuclear activities in the safeguarded state. For pre-NPT agreements, the purpose of safeguards is to ensure that the assistance provided is not used in such a way as to further any military purpose. Non-NPT agreements negotiated since the mid-1970s normally contain a commitment not to divert either to further any military purpose or to acquire nuclear weapons or other nuclear explosive devices, including peaceful nuclear explosions. Although grudgingly acknowledging this extension of NPT language (no explosive devices) into all safeguards agreements, non-NPT states have insisted that safeguards relate only to facilities and materials to which the agency has been invited to apply them, and are not intended to carry out the NPT or nuclear disarmament.

Any reference to unsafeguarded nuclear activities in non-NPT states or any attempt to draw inferences about proliferation risk or the adequacy of safeguards by the secretariat is considered by these states to be irrelevant, discriminatory, and beyond the purpose of safeguards. Given the specific and limited nature of the obligations these states have assumed, they contend (correctly) that the presence of unsafeguarded activities is fully consonant with their agreement with the agency for the application of safeguards.

Against such a background, it is not surprising that it is difficult to achieve consensus among all IAEA members on the purpose of safeguards except to say that they serve to verify compliance with safeguards agreements.

Judging Safeguards Effectiveness

Beyond the question of the purpose of safeguards lies the even more difficult problem of judging safeguards effectiveness. It is more difficult because both political and technical factors are involved that cannot be fully separated from one another. What is technically attainable may not be politically acceptable, and what is politically desirable in terms of degree of assurance may not be technically feasible. Difficulties also may arise out of misunderstandings about the intended purpose of particular measures, or out of alternative interpretations of different criteria. These problems affect evaluations of effectiveness and cause tensions with which the agency has had to contend, especially during the past decade.

Many of the relevant key concepts of safeguards and provisions of safeguards agreements were discussed in chapter 5. There it was noted that in drafting the NPT safeguards document the agency emphasized minimizing subjective judgment in favor of achieving greater precision with respect to the objectives of safeguards. The agency made an effort to separate the political purposes of safeguards (defined as assurance and deterrence) from the technical objectives. The chief technical objective of verification for NPT safeguards was defined as "the timely detection of diversion of significant quantities of nuclear material . . . and deterrence of such diversion by the risk of early detection."[50] It was further noted that the three quantifiable variables in this definition—"timely detection," "significant quantities," and "risk of early detection"—were given numerical values for use on a provisional basis, on the recommendation of the Standing Advisory Group on Safeguards Implementation (SAGSI).[51] Collectively, these values were defined as detection goals. They were recommended for use as provisional guidelines for measuring the attainment of the safeguards objective of "timely detection of diversion of significant quantities of nuclear material," and intended to serve as a management tool for the allocation of safeguards effort and for the planning and conduct of safeguards research and development. They also were seen as serving to facilitate the uniform and nondiscriminatory application of safeguards at different facilities.

Whatever the agency may have intended when it defined and used these goals, they became a prime object of attention and a target of criticism once they were in the public domain. Critics of safeguards have interpreted the detection goals not as provisional guidelines but as requirements to be met and as criteria of effectiveness. This is the interpretation given by those, including some members of the U.S. Congress, who view safeguards primarily in terms of detecting diversion rather than of providing assurance that safeguarded materials are accounted for.

Safeguards supporters have also criticized the goals, but for different reasons. They contend that detection goals, even when understood, are in some instances unattainable, and their continued use weakens confidence in safeguards generally. The detection goal of one explosive-significant quantity is considered unrealistic because in certain facilities (such as large-scale bulk handling plants) measurement error alone theoretically could exceed the proposed detection goal by an order of magnitude. From this point of view, and even if that theoretical high is not reached, the notion of allocating limited safeguards resources to try to bring diversion detection down to one significant quantity could cause a very severe drain on those resources and still not give absolute assurance over a long period of time that a significant quantity of material has not been diverted without detection, since there will always be some measurement error.

The timeliness goal presents similar problems for weapons-usable materials such as separated plutonium. Those who claim that this goal too is unreasonably stringent point out that in virtually all situations short of an inspector's actually observing an unauthorized removal of nuclear material under safeguards, all that he can detect and report are anomalies and not diversions in progress. Anomalies are "unusual observable conditions which might occur in the event of a diversion," such as damaged seals, excessive quantities of material unaccounted for, or inconsistencies in records.[52] They lead to investigations that entail reviews of records and reports, requests to the state for further information, and so on, a procedure that can take weeks or even months after the inspection interval. Partisans of this view also contend that except under the most unusual circumstances it is unlikely that diverted material could be converted into a nuclear explosive device, let alone deployed and used against another state in the time periods in question.[53]

This has led some supporters of safeguards to recommend altering the numerical values associated with the different variables to conform more closely with the actual capabilities of the safeguards system. Others, however, have opposed this approach, maintaining that marginal adjustment in values would have only marginal impact on perceived safeguards effectiveness, while significant changes could so alter the anticipated levels of detection (for example, 50 or 100 kilograms of plutonium at a chemical reprocessing instead of 8 kilograms, which is the current significant quantity definition for plutonium) as to undermine general confidence in the safeguards system.[54]

For its part, the IAEA has distinguished between detection goals (ultimate objectives) and inspection goals, which are basically detection goals adjusted to what is feasible and presumably attainable, although they are not always achieved in practice. Over time, however, the number of cases in which the agency has fully attained its inspection goals has increased substantially, as has the number of cases where inspection goals have been attained with respect to plutonium and highly enriched uranium outside reactor cores.[55] In addition, the IAEA has introduced the accountancy verification goal, which defines the minimum quantity of nuclear material whose diversion could be detected with the required detection probability by using only material accounting measures. For large bulk handling facilities, this goal tends to exceed one significant quantity and constitutes the agency's recognition of the unfeasibility of achieving detection goals at some of these facilities.[56]

Public awareness of the uncertainties surrounding safeguards goals has led critics to question the credibility of IAEA safeguards. Based on the known technical limitations of those safeguards, some argue that certain nuclear materials and facilities—notably plutonium and reprocessing plants—are "inherently unsafeguardable" and that the agency and its safeguards system will be irreparably damaged if it is called upon to perform the allegedly impossible task of "safeguarding the unsafeguardable."[57] Therefore, they advocate the termination, or at least sharp curtailment, of sensitive fuel-cycle activities such as reprocessing, and discount the view that IAEA detection goals should be accepted for what they are—goals toward which the IAEA should strive, guides for rational allocation for safeguards resources, and for safeguards research and development.

It has been suggested that an alternative measure of safeguards effectiveness is the amount of inspection effort expended.[58] The relevant concepts—maximum and actual routine inspection effort (MRIE and ARIE)—were introduced in chapter 5. As noted there, the NPT safeguards document specifies the maximum routine inspection effort, but the actual effort is the result of negotiation between the agency and the state, and takes into account the presence of containment, surveillance, and other factors. Typically, ARIE amounts to about one-quarter to one-third of MRIE. In addition, the agency annually determines the planned actual routine inspection effort (PLARIE) it will deploy at each inspected facility, taking into consideration the operator's predicted program of activities. PLARIE is significantly lower than what theoretically could be applied, and often lower than the agreed actual routine inspection effort, ARIE; even so, it is not usually met.

Criticism of PLARIE comes from two directions. Some non-nuclear weapon states with large nuclear programs that are already subject to a substantial safeguards effort are reluctant to accept the idea that PLARIE figures alone could serve as an appropriate measure of safeguards effectiveness. In their view, such agreements would only lead to increased safeguards activity in the least proliferation-prone countries and, parenthetically, greater pressure for larger safeguards budgets. A different criticism is that since inspection effort levels are quite far from the original detection goals, an effort to substitute them as a basis for judging effectiveness could lead to a sense of sleight of hand, attempting to demonstrate that effectiveness is better than assumed. From this perspective, inspection effort may be a measure of something, but not of safeguards effectiveness. From both points of view, the use of planned routine inspection effort to measure safeguards effectiveness, like the safeguards goals as they now stand, invites misunderstanding, overexpectation, and inevitable criticism and loss of confidence.

Safeguards effectiveness remains a conceptual and a practical problem:

- How should it be defined?
- What factor (or combination of factors) should be the focus?
- Can certain limitations be overcome only as a result of institutional changes (multinational arrangements or the suspension of certain activities), or would a more efficient use of a larger resource base solve the problem?
- Is the core of the problem that the purpose of safeguarding (verification or detection) is ill defined, hence expectations are inappropriate?

Troublesome questions remain as to whether safeguards performance should be judged in numerical terms or by political confidence in the system, and how the relationship between numbers and confidence should be defined. An imperfect system that enjoys high political credibility probably would be preferable to a perfect system that does not. If, of course, the concern lies beyond either assurance or detection of diversion and extends to prevention of abrogation by a state of its pledge not to use its peaceful nuclear program for prohibited military purposes, then the problem falls *outside* the realm of international safeguards and squarely in the lap of the nonproliferation regime as a whole.

Safeguards Implementation

Beyond the largely conceptual issues of defining the purpose of safeguards and identifying a consensual basis for evaluating their effectiveness are other issues that raise questions about safeguards credibility, but do so in relation to more concrete factors and events. Some of them had been present for a long time, but became more prominent and more urgent when former IAEA inspectors challenged the adequacy of agency safeguards to deal with sensitive nuclear materials in testimony before congressional hearings held in the wake of the Israeli attack on the Iraqi reactor in 1981. The issues most often raised are discussed here.[59]

A Bill of Particulars

The issues involved in safeguards implementation embrace political, technical, and management concerns.[60] Political issues relate to restraints in safeguards implementation that reflect the status of safeguards as a novel institution that intrudes into the sovereignty of states that for centuries have jealously guarded that prerogative. These restraints (real or alleged) include the following:

- Limitations on where safeguards can be carried out. Inspectors are not free to search for materials or facilities that have not been declared by the safeguarded state. This applies even though a state may have obligated itself to place all of its nuclear activities under safeguards.
- Limitations on inspector access or safeguards activity. These limitations apply to a facility in a state under NPT-type safeguards before material has been introduced into the facility or when nuclear material is no longer present.[61]
- Limitations with respect to the designation of inspectors. States retain the right to determine whether or not to admit particular foreign nationals. Practices have emerged wherein states have declared in advance that certain classes of persons are not acceptable, such as inspectors of a particular nationality, or those from states that do not themselves accept inspection, or those from states that refuse to accept nationals from the state being consulted about an inspector designation. Some states also have insisted on a limit on the number

of inspectors designated for conducting safeguards in their jurisdiction (that is, the total number of inspectors that can constitute the pool of acceptable inspectors). These practices are viewed as tantamount to veto power and as a real impediment to efficient safeguarding.

• Limitations on publicizing information upon which states and the relevant public can make considered judgments on the credibility and value of the safeguards system.

• The lack of political spine on the part of inspectors or the agency to promptly sound the alarm if they are not able to affirm peaceful use or if they detect a serious anomaly that cannot be reconciled.

Technical issues relate to the inspectorate, equipment and instrumentation, and the adequacy of existing methods and procedures to safeguard effectively all nuclear fuel cycle activities. These include:

• The quality of the inspectorate in terms of their background, degree of training to deal with safeguards responsibilities, and their professionalism. Implicit in the latter is the claim that inspectors are "very responsive to concerns of the countries (they) inspect,"[62] a point that is very close to the question of willingness to sound an alarm where material cannot be adequately accounted for.

• The adequacy of instrumentation and equipment used for measurement, containment, and surveillance. Questions have been raised about the quality of measurement instrumentation as well as about the quality and reliability of cameras, seals, and other complementary measures. The general thrust of criticism here is that safeguards methods and procedures are not always sufficient to achieve effective safeguarding.[63]

• The agency's inability to verify through materials accountancy alone whether significant quantities of plutonium as may be unaccounted for at reprocessing facilities represent measurement uncertainties or possible diversion of nuclear material. This focuses on a central concern of many in the nonproliferation community, namely the ability of safeguards to deal with high-throughput bulk-handling facilities that handle weapons-usable material. Here again, the notion is that because it is unfeasible to adequately safeguard reprocessing plants, the further establishment of such facilities should be discouraged.

Management issues relate to the smoothness with which inspection operations are carried out, including the examination and evaluation of inspection data, the degree to which evaluation procedures are standardized across the different safeguards operating divisions within the directorate of safeguards, and similar matters. Perhaps the main area of concern is the treatment of anomalous situations, several hundred of which arise each year. How these situations are resolved is not well understood among those whose confidence in safeguards ultimately can determine the value of safeguards and their role in nonproliferation.

Assessing the Charges

Standing alone, this list of particulars is somewhat awesome. How is it to be evaluated?[64] The short answer to the political criticisms could be that the activities in which the IAEA is engaged are novel and touch the very heart of sovereignty and therefore are to be expected: international organizations have their limits; they are creatures of the states that created them and lack independent authority. But this reply fails to adequately acknowledge the security relevance of safeguards and the general importance of nonproliferation. A closer look indicates that while there are indeed limitations and weaknesses of the kind described, in many instances they are less daunting than they appear, and they are in any event the subject of continuing and often intensive corrective efforts.

In some cases, it is less a question of whether the agency has particular rights than whether and to what extent it will exercise them. In others, it is a situation of real limitations to which adjustments must be made. Thus, while it is true that the IAEA does not have unlimited access even within NPT states, and lacks authority to seek out undeclared facilities or material, it is also true that in the case of NPT non-nuclear weapon states, the agency has the right and obligation to apply safeguards to all peaceful nuclear activities, and for its part the state is obliged to declare all nuclear material. A state that fails to do so is in default of its commitments under the treaty and of its safeguards agreement. Two other factors are relevant here: (1) that it is not unreasonable to assume that states taking on commitments intend to honor them and have a stake in providing all information necessary to secure a clean bill of health from the independent inspecting authority; and (2) that in a very large number of states the agency already has a running record of nuclear activity and can, therefore, have substantial confidence in its findings. Furthermore, inspections may reveal discrepancies that suggest the presence of undeclared materials or facilities, and the agency has the right under the safeguards document to call for further information and to request more extensive inspection if anomalous situations cannot otherwise be resolved. A state's reluctance to cooperate can lead the director general to report this to the board. While not detecting a diversion, this whole process essentially enables safeguards to fulfill their responsibility—as defined in INFCIRC/153—to verify that all nuclear materials are accounted for, or to serve notice that such verification cannot be achieved. The question is both one of rights and of the agency's determination and will to exercise them.

This raises the question of political will. The possibility cannot be ignored that an anomalous situation laced with ambiguities and involving a less than cooperative state could lead the secretariat to defer, even indefinitely, informing the board. But the record to date seems to indicate a different result. Between 1981 and 1983, the agency made the determination that it was not in a position to verify that no diversion had taken place at several facilities subject to its safeguards. This finding of non-verification demonstrated that the IAEA was capable of reaching such a conclusion and was prepared to bring the issue before the board of governors. With the support of the board, the agency was able to renegotiate the safeguards arrangements and to bring the facilities in question under effective safeguards. The task was not easy, nor was it achieved quickly, but it was done. In another case in 1984, an NPT state had failed to notify the agency of the export of a quantity of depleted uranium. This created an anomalous situation that was picked up by the safeguards system and subsequently judged to be an accounting anomaly. This incident also was brought to the attention of the board.

Access to facilities into which nuclear material subject to safeguards has not been introduced raises an issue that has been the subject of frequent debate. Some argue that INFCIRC/153 safeguards apply only to nuclear material; others contend that pressing this argument only leads to legalistic interpretations that overlook the fundamental fact that in practice the application of safeguards involves the facilities intimately because they normally cannot be separated from the material they contain. Safeguards may apply only to materials, but materials are located at facilities.[65] Moreover, the agency has the right to request special inspections involving access to locations in addition to those open to routine inspections if it is of the view that the information it is receiving from the safeguarded state "is not adequate for the Agency to fulfill its responsibilities. . . ."[66]

Designation of inspectors has posed problems for the agency because it has undercut efforts to improve safeguards efficiency, which for many states goes hand in glove with safeguards effectiveness.[67] On the other hand, there has been some improvement in recent years. The increasing size of the inspectorate has provided an added degree of flexibility by deepening the pool from which inspectors can be drawn. Whereas in 1980 there were only about 750 approved designations in 51 states, in 1985 the IAEA had secured 1,500 designations in 56 states. This is an ongoing problem and the agency is currently seeking agreement that all inspectors assigned to one of the three operating divisions in the Department of Safeguards would be accredited to all states covered by that division. This has not yet become a reality, however. Admittedly, the designation issue has not yet resulted in a crisis of confidence in the agency's ability to objectively carry out its safeguards responsibilities. But that could come at any moment, and now may be the time for the director general to impress upon the board of governors the urgency of resolving the matter of inspector designation and averting the risk that limitations on designation might lead to questions about the credibility of safeguards.

In addition to improving the situation for inspector designation, the agency has taken steps to overcome the problem of securing visas for inspectors. In some cases, visas are required for every visit, thus rendering an unannounced inspection virtually impossible; in others, visas are not required. The agency already has established field offices in Japan and Canada, thus eliminating the visa problem in two states with major nuclear programs. And it is working toward multiple-visa arrangements in other parts of the world.

The publicizing of safeguards involves two audiences. One is the board of governors, which has overall responsibility for the agency. The other is the relevant public in the member states. In 1977, at the prodding of the United States, the agency began to prepare an annual Safeguards Implementation Report (SIR) to inform the board of developments, problems, and progress in the implementation of safeguards. These are extensive documents that are constantly being revised in format to provide readable, comprehensive, and meaningful reports. Traditionally, the information is provided in categories, and states and facilities are not named, in keeping with the effort to remain sensitive to sovereign concerns. The general public is kept informed primarily through the Annual Report, which reproduces the first (summary) part of the SIR and provides a general overview of safeguards activity. The detailed report, however, is not published by the agency or released by its members.

The main tension that arises in publicizing safeguards activities and results is that between transparency and confidentiality. Transparency has been a goal not only of safeguards critics but of many supporters as well who are persuaded that since the ultimate credibility of safeguards depends on perceptions of their effectiveness, there must be enough information in the public domain to permit informed judgments. Confidentiality, on the other hand, is the byword of sovereignty, and many states, whether for reasons of seeking to protect proprietary commercial information, or as a matter of general principle (or even because of inadequacies in the quality of their national accounting and control systems) seek to ensure that any move toward transparency does not violate the strong admonition in both safeguards documents to protect information obtained by the agency in the course of applying safeguards.

As for the technical issues, an evaluation of the claims reveals a more positive and encouraging picture than might be assumed. As for the inspectorate, a lifelong participant in the international nuclear arena recently noted that "Agency safeguards personnel, including the inspectors, are an impressive group, demonstrating technical capability, political judgment and seriousness of purpose in the vast majority of cases" and that "this assessment is widely shared by officials of safeguarded facilities themselves."[68] This is, of course, a subjective judgment, but it is quite in keeping with the consensus view about the overall safeguards system expressed by the participants at the Third NPT Review Conference that already has been mentioned and that will be returned to shortly.

For its part, the IAEA has established a training section devoted exclusively to improving the training of inspectors; training manuals have been developed, and an intensive training program has been carried out, especially in the past five years. This includes field exercises at nuclear plants for all new inspectors, advanced courses in inspection procedures, and refresher courses at regular intervals for seasoned inspectors. This does not mean that problems do not exist. For example, as might be expected in any organization having both operational and evaluation divisions, there is a conflict of interest between operators whose performance is measured in terms of production, and development personnel (including analysts who design safeguards approaches) whose efforts lead to changes and hence to disruption of production. This undoubtedly will continue to have an impact on strategic choices regarding research and development, selection of inspection instrumentation for use in the field, and distribution between field and home base insofar as evaluation activities are concerned.[69]

The quality of equipment and the adequacy of instrumentation for safeguards is a continuing problem that has been the subject of a good deal of agency research and development. During the congressional hearings following the Israeli attack in June 1981, some emphasis was placed on the failure rate of film cameras, the alleged poor quality and easy duplication of agency seals, and the unreliability of measurement instruments that, because of their role in safeguarding of material accountancy, played a very central role.[70]

With the assistance of a number of member states (for example, the United States through its Program on Technical Assistance for Safeguards (POTAS) and counterpart programs in a half-dozen other advanced nuclear states), the agency has significantly upgraded safeguards equipment during the past five years. The failure rate for cameras, for example, is only one-third the characteristic rate for 1980, and much greater emphasis is being given to closed-circuit television to supplement or replace surveillance cameras or both. Fibre-optic seals are playing a progressively more important role, supplementing cup and wire seals, and research and development is being conducted on ultrasonic seal devices. Canada provided the necessary research and development to develop substantially improved safeguards at on-load fuel reactors of the CANDU type, which now exist in Canada, Korea, India, and Argentina; Argentina has been cooperating with the agency in improving the quality of safeguards at this type reactor. As in the case of the inspectorate, this does not "solve" the equipment problem, but it narrows an important gap and serves to upgrade safeguards.

In general, technical safeguards capabilities at facilities such as light water power reactors (which account for 80 percent of power reactors under safeguards), small power reactors, and (with the Canadian fuel bundle counter device mentioned above now in place) in heavy water power reactors are adequate to the safeguarding task. On the other hand, bulk handling facilities, as discussed earlier, present their own set of less tractable problems. Experience with centrifuge enrichment facilities (the only type of enrichment plant now under safeguards) is still too limited to provide a data base for drawing meaningful judgments. But it is significant that after several years of study and discussion, the IAEA, EURATOM, and six countries reached agreement on the safeguards regime to be followed in such facilities, including the use of a limited number of unannounced inspections.

Reprocessing plants are the other major concern in the nonproliferation community (along with fuel fabrication plants that produce plutonium/uranium mixed oxide fuels). A number of advanced nuclear states have been cooperating with the IAEA in seeking to develop concepts and procedures for effectively safeguarding plutonium facilities. One approach seeks to introduce "near realtime accounting" into national material accounting systems in a manner suitable to independent international verification. When functioning as intended, this approach presumably would enable a safeguarding organization to verify operations within a very short time span through frequent measurements at short intervals. Notwithstanding this effort to achieve higher quality and more timely information, it is clear that plutonium presents a special case, and that in a material accountancy situation, even the most precise and timely inspection arrangements (including continuous inspection arrangements) may leave more to be desired than is comfortable. This impels the search for additional measures such as multinational approaches to plutonium separation and the development of some form of international plutonium storage arrangements. While some years of effort have been put into the latter, relatively little progress has been made, due to fundamental differences over such concepts as "end-use control" wherein the storage authority would be able to verify that the stored plutonium was being removed for authorized and legitimate purposes.

Endorsement of the NPT Review Conference

The director general of the IAEA frequently has emphasized the value of safeguards as an international confidence-building measure. The preceding discussion of problem areas and progress in the arena of safeguards may give reason to wonder about reaching a conclusion that safeguards foster such confidence. Perhaps the most authoritative judgment was that rendered by the eighty-six states party to the NPT who participated in the Third NPT Review Conference in August and September 1985. The results of that conference are important and germane to our discussion. The conference report expressed:

> the conviction that IAEA safeguards provide assurance that states are complying with their undertakings and assist states in demonstrating this compliance. They thereby promote further confidence among states and . . . help to strengthen their collective security. IAEA safeguards play a key role in preventing the proliferation of nuclear weapons and other nuclear explosive devices.

Furthermore,

> The Conference notes with satisfaction the improvement of IAEA safeguards which has enabled it to continue to apply safeguards effectively during a period of rapid growth in the number of safeguarded facilities.

> It also notes that IAEA safeguards approaches are capable of adequately dealing with facilities under safeguards. . . .[71]

A concurring endorsement came in the November 1985 summit meeting between President Reagan and General Secretary Gorbachev.[72] Of course, these statements do not solve the many problems confronting effective safeguards today; but they do provide the basis upon which to carry forward the long-standing objective of making nuclear energy feasible in a world of nation states. And most importantly, they reflect the kind of positive and forward-looking attitude that is absolutely essential for there to be any success in constructive strengthening of the nuclear nonproliferation regime.

DIFFERING AGENDAS FOR NPT AND NON-NPT MEMBERS

Since 1970, when the Treaty on the Non-Proliferation of Nuclear Weapons came into force, the IAEA has been responsible for administering the safeguards required by that treaty. Most of the agency's members have become parties to the NPT, but a number of member states have chosen not to undertake the NPT commitments. Several of the latter—including India, Argentina, Brazil, and Pakistan—have become increasingly critical of what they perceive to be a steady pressure by the secretariat and some member states toward conforming all agency safeguards activities to the standards of the NPT. Among other drawbacks, they see this as entailing the functional equivalent of imposing full-scope safeguards without any voluntary agreement to do so.

This trend, in their view, distorts the application of the agency statute under which safeguards are a purely voluntary arrangement and wherein states, when they do accept them have the option, depending upon the source of the safeguards obligation, to accept safeguards on imported materials and items only or on all of their nuclear activities. These nations also regard this trend as a challenge to the legitimacy of their own more limited concept of safeguards, requiring them to defend what appears to others an unorthodox point of view. They see this occurring not only in the occasional suggestions to bring the non-NPT and NPT safeguards documents into line with one another, but in other more subtle ways as well, for example in efforts by the secretariat to achieve that result by creating new safeguards rights and precedents through the medium of technical improvements and updating of agreements to take advantage of advances in the state of the art of safeguarding.

The more the IAEA appears to act as agent for the NPT, the greater the tension becomes. In an effort to come to grips with this problem, and to ameliorate the tension it creates, Director General Blix in 1984 said:

> There might sometimes be a tendency to view Agency safeguards as a kind of appendage to the Treaty on the Non-Proliferation of Nuclear Weapons (NPT). . . . The Treaty had in fact enabled a very large number of non-nuclear weapon states, larger than the number of Agency member states, solemnly to commit themselves to foregoing nuclear weapons and to ensure that their compliance was verified through safeguards. . . . However, notwithstanding the very great importance of the NPT, it should be borne in mind that IAEA safeguards were based directly on the Agency's Statute. NPT made use of the Agency's safeguards, as did the Tlatelolco Treaty. . . . [but] It should not be overlooked . . . that a State so wishing could still, through a bilateral agreement with the Agency, commit itself to safeguards verification that all its present and future nuclear activities were for peaceful purposes.[73]

Even though this statement implied a preference for full-scope rather than partial safeguards, some of the concerned non-NPT states have viewed it as reaffirmation of the priority of the statute in agency affairs.[74] The largely political and symbolic character of this problem can be seen from the fact that current differences in actual application of the two safeguards documents are insignificant.

Objections to Safeguards Implementation

As discussed earlier, since 1977, the agency has prepared a Safeguards Implementation Report (SIR) each year, to provide better information to member states about safeguards implementation and to improve the basis upon which members can assess the quality and effectiveness of IAEA safeguards. The SIR contains the data base upon which the secretariat and subsequently the board of governors draw a conclusion as to whether all safeguarded nuclear material remained in peaceful nuclear activity or was adequately accounted for otherwise. The reports identify problems that exist but do not name problem states or facilities, because their presentation is based on the principle of anonymity to protect the confidentiality of safeguards information and specific provisions of safeguards agreements.

Non-NPT members nevertheless have taken issue with the manner in which information is presented, most particularly because it distinguished between safeguards carried out under the NPT and non-NPT arrangements and provides information on unsafeguarded nuclear facilities, materials, and activities in non-nuclear weapon states. Non-NPT members argue that it is not the agency's business to say anything about activities that are outside its responsibility, and that by so doing the agency discriminates against some IAEA member states in a manner contrary to its statute. Furthermore, they contend that even mentioning unsafeguarded facilities implies questionable behavior on the part of the state involved, thereby impugning its motives and integrity. The burden of this argument is that the purpose of safeguards is to verify compliance with safeguards agreements and not to raise questions about nonproliferation or other such matters; that the secretariat should indicate possible cases of noncompliance with agreements but not cast doubt on compliance by commenting on the existence or nonexistence of facilities or materials not subject to safeguards.

IAEA Articulation of External Agreements

Non-NPT members also complain about the agency's growing role as a vehicle for articulation of NPT and other nonproliferation regime agreements reached by states outside the framework of the IAEA. They criticize, for example, the agency's publication of the London supplier guidelines and its adoption as guidelines of principles that are not, in their view, consistent with the agency's statute.[75]

An important case in point was the decision by the board of governors in February 1979 to approve new technical assistance guidelines, which stated the principle that the peaceful uses of atomic energy exclude "research on, or development, testing or manufacturing of a nuclear explosive device."[76] This removed peaceful nuclear explosions from the ambit of acceptable nuclear activity, which, while fully consistent with the NPT, went beyond the IAEA statute. It also reflected language incorporated into more recent safeguards agreements concluded between the agency and such non-NPT member states as Argentina and Spain.[77] These agreements provide that none of the items subject to the agreement "shall be used for the manufacture of any nuclear weapon or to further any other military purpose or for the manufacture of *any other nuclear explosive device*" [emphasis supplied].[78]

Although Argentina had negotiated a safeguards agreement containing this language, it strongly protested its introduction as a general criterion. Argentina charged that this was foreign to the agency's statute and was a surrogate way of enforcing on the entire membership principles agreed to by only some of the agency's members.[79] Brazil and India have adopted a similar position.[80] Both India and Argentina continue to argue the legitimacy of peaceful nuclear explosives. They have refrained from making any requests for technical assistance since the adoption of the guidelines by the board. More generally, these and other non-NPT members take the view that the IAEA is heading toward making achievement of nonproliferation objectives its primary concern, even at the risk of hampering the development—let alone, promotion—of peaceful nuclear activities. This view is also shared by some NPT developing states.

Another example of the problem discussed here can be found in the field of technical assistance. As discussed below, the agency provides funding for projects deemed to be technically qualified for assistance from the resources made available by member states in the form of voluntary contributions. Technically qualified projects for which funding is unavailable are incorporated in a footnote, the so-called "footnote A," of the Report of the Technical Assistance and Cooperation Committee to the Board of Governors. Supplier states reviewing the report traditionally provide additional funds to support some projects included in this footnote. Most of the key supplier states are parties to the NPT, and the majority of them have been giving preferential treatment in funding "footnote A" projects to developing countries that are parties to the NPT. Non-NPT member states individually and as a group have protested the use of the agency as an instrument for applying an external treaty that results in discrimination among agency members as contrary to the IAEA's statute.[81]

The main concern of the non-NPT states is to preserve the legitimacy of their own peaceful nuclear activities, even if it means arguing for the legitimacy of peaceful nuclear explosives (which they may not even intend to pursue) and against full-scope safeguards. They seek affirmation of that legitimacy by agency institutions such as the board, if not by positive acts, then at least by avoidance of actions that put them on the defensive. They seek to maintain a definition of nuclear development that precedes the NPT, and to ensure that this remains a part of the IAEA's credo.

Efforts to Mobilize the Group of 77

To achieve these objectives, the non-NPT members seek to mobilize the G-77, which, as we have seen, contains some states with little current interest in nuclear energy and, therefore, no particular stake in protecting the organization against political buffeting from confrontations between NPT and non-NPT member states. These states accept whatever benefits they can derive from technical assistance, and they are amenable followers of the articulate, and sometimes aggressive, leadership of countries such as India, which have well-defined interests and clearly defined agendas and are fully prepared to trade on Third World solidarity to achieve their objectives. They tend to see things more in terms of North-South conflicts than do others. Moderate members of the G-77 who also are NPT parties, but who are interested in seeing the agency serve as a channel of reward for their treaty adherence, are caught between their NPT status and their sense of solidarity with the Group of 77. Thus far, they have been unable to abate the confrontation over these issues, or to establish unequivocal leadership aimed at constructive reinforcement of the agency. There is considerable concern both in and outside the IAEA as to whether these differences can be contained to avoid undermining the institution.

Reconciling NPT and Non-NPT Members

The challenge facing the principal IAEA members today is that of reconciling two groups of states that are divided by their participation in a treaty that reflects values and commitments not uniformly shared by the entire membership. There is no way to effectively divorce the NPT from the IAEA, and there can be no acquiescence by the majority in the fundamentalist values of the non-NPT members. For the agency to continue as a viable and effective institution, a way of reconciling these conflicting relationships must be found. This is one of a number of important challenges for U.S. leadership in the agency.

BALANCING SAFEGUARDS AND TECHNICAL ASSISTANCE

Historically, the advanced nuclear states have dominated the IAEA. With their nationals occupying most of the principal posts in the secretariat, they have largely shaped the agenda for the board of governors, set the programs, and defined the rules of the game whereby the agency operates. Although their influence is more tenuous today, for a long time this dominance by the big powers was accepted by most of the membership, which felt their interests were reasonably represented and that the agency was generally responsive to their principal concerns.

Of course, in the earlier years the number of countries with nuclear power programs was smaller, as was the kind and amount of assistance that most of the member states could reasonably absorb. And until the agency was made responsible for implementing the safeguards provisions of the NPT, resource demands for this activity remained limited and modest in growth. From 1958 to 1967, for example, agency expenditures for safeguards totaled slightly over $3 million, while more than $23 million was spent on all of the other program activities. The NPT made a difference. Between 1978 and 1982, the agency spent $109 million on safeguards and $116.7 million for all other activities.[82]

Some Statistical Comparisons

During the twenty-five years from 1957 to 1982, the IAEA spent over $213 million for nonsafeguards activities, including providing for training courses, experts, equipment, research contracts, fellowships, publications, advisory missions, and professional meetings. Technical assistance activities were modest during the first fifteen years, totaling $43.7 million, but this accelerated in the following decade to reach a total of $169.5 million between 1973 and 1982. Safeguards funds in the same twenty-five-year period totaled $147.2 million. Funding for safeguards, however, had increased at twice the rate for technical assistance over the last ten years of that period and almost equaled the total expenditures for all other programs for 1978–1982.

In the past several years, these trends have been somewhat modified because of substantial increases in the resources being made available through voluntary contributions to the Technical Assistance and Cooperation Fund. In 1981, for example, the target for contributions to that fund was $13 million (compared with $8.5 million in 1979), but by 1985 it had doubled to $26 million. The target for 1986 was $30 million, and for 1987, $34 million. As a general rule, more than 90 percent of pledges made are fulfilled. Of course, technical cooperation funds are not limited to contributions to the fund, but also include extra-budgetary financial resources, assistance in kind, and UNDP funds.

Safeguards funding by way of comparison was $26.3 million in 1981, $33.6 million in 1983, $36.6 million in 1984, and $39.7 million in 1985. Of course, well over 90 percent of safeguards expenditures are provided through the assessed budget and do not depend on voluntary contributions, although such extra-budgetary resources are made available to the agency by a number of advanced nuclear states.[83] To put the 1984 figure into perspective, it may be noted that for the same year the police force budget (admittedly involving activities beyond the scope of the IAEA) for Portland, Oregon, a modest-sized American city (population 372,892), exceeded $39 million.

The agency's technical cooperation and assistance activities, while modest by some standards, are nevertheless significant for the international nuclear regime. Its safeguards services are the condition sine qua non for any significant international nuclear cooperation and commerce. Agency allocations for these two major functions, which are funded on different bases, are nevertheless a source of friction among IAEA members.

Log-Rolling by Developing Countries

Developing country member states have not been quiescent. Historically, one of the first resolutions passed by the general conference called for priority on activities of direct relevance to developing countries, and this caused some changes in the agency's initial program.[84] The developing countries also lobbied hard and successfully against resistance from many advanced nuclear states (and the advice of the director general's Scientific Advisory Committee) to establish an International Center for Theoretical Physics at Trieste, and later a regional radioisotope center in Cairo.[85] They also succeeded in altering the distribution of agency research contracts by winning agreement that preference would be given to developing countries if the work accomplished there would be basically equivalent to work that could be carried out in the laboratories of an advanced nuclear state. Nearly two-thirds of almost 2,000 contracts awarded between 1958 and 1982 went to developing country member states as a result.[86] In 1966, following an internal review of agency activities that was initiated by the developing countries, the general conference adopted a resolution allowing the agency to supply equipment for technical assistance purposes independent of any expert services.[87] This important decision, at variance with normal UN practice, set the stage for a pattern of IAEA technical assistance that included a much larger share of hardware than is common in international technical cooperation.[88] In earlier years, equipment normally would be accompanied by an expert to facilitate training and use for supplied equipment. Commencing in 1985, many regular and "footnote A" projects are once again being assigned a training component, apparently in an effort to improve the effectiveness of the assistance being provided.

Third World Efforts to Name a Director General

Similarly, on matters of representation, developing countries periodically have sought greater control over administration of the agency. Although covered in the discussion of politicization, this subject deserves some further comment because of its relationship to attempts by Third World countries to exert influence over the direction and emphasis of IAEA activities. When Sterling Cole, the first director general, retired in 1961, the developing countries nominated Sudjawro of Indonesia for the post, arguing that they were inadequately represented in the top echelons of the agency and that the best way to rectify the situation would be to install a Third World national at the head of the secretariat. They were unsuccessful.

Twenty years later, following the retirement of Sigvard Eklund, a much larger and more aggressive group of Third World nations, after again failing to win the necessary two-thirds majority for their candidate, tried unsuccessfully to secure adoption of an arrangement whereby the director general's post would rotate between advanced and developing nations. The compromise resolution finally reached nevertheless provides that in the future, candidates from all regions should be given serious consideration. So it is altogether possible that when Hans Blix retires, a Third World candidate will be elected to the post.[89]

A Bias Toward Safeguards

The issue of balance has been stimulated by what Third World countries see as an increasing bias by the agency toward safeguards at the expense of technical assistance. This view is widely shared across the full spectrum of developing country member states, from those that are well advanced to those with only rudimentary nuclear programs and interests; and by those that are NPT parties as well as those that are not.[90] Non-NPT states have used the perception of imbalance to generate sentiment against significant increases in agency safeguards activity. The problem has somewhat abated recently in light of the significant increase noted earlier in resources available for the agency's technical cooperation program. But it lingers just beneath the surface and is defined not only by the amount of assistance available at any given time, but also by the degree of certainty of its availability, as will be discussed below.

The argument over balance reflects the comparatively sharp increase of resources allocated to safeguards as compared with technical assistance in the mid-1970s to early 1980s. Counting resources can be a tricky business. The representative of Pakistan speaking at the 1984 general conference, for example, noted that since 1970 the agency's safeguards budget had increased by a factor of 27 (from $1.23 million to $33.7 million), while allocations to technical assistance for the same period of time had increased only by a factor of 9 (from $2.7 million to $23.5 million).[91] If, however, one includes in technical assistance allocations both the voluntary cash contributions that go into the IAEA's Technical Assistance and Cooperation Fund, and other extra-budgetary contributions, in-kind gifts, and UNDP resources, the 1984 total for technical assistance rises to approximately $36 million—a sixteen-fold increase.

An important related consideration is that almost from the inception of rising safeguards costs, a safeguards-financing formula has been in effect. This formula relieves developing countries from the burden they would have to shoulder if they were to pay the normal full contribution to safeguards. At the same time, it is intended to implement the principle that since safeguards benefit the entire international community, not only by making global nuclear commerce possible, but also by contributing to international security, all states should participate in their financing. Under this arrangement, the nuclear weapon states pay more than 50 percent of the safeguards budget and 36 member states pay more than 98 percent of safeguards costs. The entire Third World membership of the IAEA collectively contributes less than 2 percent of the safeguards budget. The formula in question was first approved in 1971, revised in 1976, and has continued since then. In 1984, the general conference asked the board of governors to review the arrangement and make a recommendation that would support a long-term solution to the question of safeguards financing, but thus far no consensus has emerged around any new formulation.[92]

Nonetheless, there remains a significant difference between the rate at which the two budgets have grown (a sixteen-fold versus a twenty-seven-fold increase), especially if one takes the view, as Third

World countries do, that at the very least, IAEA promotional activities deserve to be treated on a par with safeguards.[93] The existence of this financial gap, and what it implies in terms of priorities within the agency, is predictably of concern to Third World members. That concern generates Third World pressure for more technical assistance, including more resources. It also fuels efforts, discussed earlier, to restructure the board of governors by increasing Third World representation.

Reliable Funding for Technical Assistance

Perhaps even more than levels and trends in support for technical assistance, developing countries have been concerned about the reliability of funding for this purpose. While safeguards are included in the assessed (regular) budget, technical assistance depends almost entirely on voluntary contributions that are not part of the regular budget, although they have been stable over the long term. There have been periodic pressures to finance technical assistance through the regular budget. In 1981, the general conference adopted a resolution calling for funding of technical assistance through the regular budget or by other "comparably predictable and assured means."[94] Donor states have strongly resisted the idea of integrating technical assistance funding into the regular budget. They argue that this would be quite inconsistent with the usual practice in international organizations, but their unstated reason almost certainly is the politics that this kind of measure would bring to the board of governors, and the likely effect of such politics on the agency's ability to carry out its activities, especially safeguards whose financing then would be formally linked to technical assistance funding. Furthermore, this change could be achieved only by statutory amendment, which creates its own set of problems in the minds of many member states.[95]

"Comparable means" however, have been devised. The agency has informally adopted what is called an indicative planning approach, whereby voluntary contribution targets are agreed upon by major contributors for periods of three years at a time. The approach has had remarkable success and earned the praise and support of the vast majority of its membership for providing substantial resources for technical assistance on what has thus far been a reliable and assured basis. Thus, from 1980 to 1985, when the regular budget increased by 55 percent, technical assistance funding increased by nearly 250 percent, based on the indicative planning figures approach. Whereas the target for technical assistance funding before

1970 never exceeded $2 million, it reached $10 million by 1980, $22 million by 1984, and $30 million in 1986. The percentage increases have been impressive, ranging from 12 percent now set for 1987, 1988, and 1989, to over 23 percent at the initial stages of indicative planning. While the percentage increases have diminished in recent years, the absolute increases have been rising—from between $2 and $3 million in the initial stages to between $3.5 and $4.0 million for the period between 1984 and 1989. And in the process it has contributed to agency stability by avoiding the tensions and attendant political controversy characteristic of annual budget negotiations, as well as the risk that discussions of the budgetary support level for safeguards would be linked to the level of technical assistance being provided. To the contrary, it has been possible to separate the two and thereby to contribute, if modestly, to diminishing politicization.

Ironically, this has come at a time when zero-growth budgeting has been applied to international organizations generally, including the IAEA. Although many of the agency's program activities have felt the pinch of budget stringency, technical assistance, as demonstrated, actually has taken some of its most impressive leaps. Voluntary funding, in other words, has allowed the agency to increase technical assistance faster than any other activity while maintaining the appearance of a zero-growth budget. Recognizing this, a number of the developing states have endorsed continuation of the concept of indicative planning and have given enthusiastic endorsements to the method and its output at meetings of the general conference.[96] While continuing to stress the importance of both augmenting technical assistance and securing assured and predictable financing, these states have come to recognize that changes can incur risks especially at a time, like now, when zero growth budgeting has become a watchword for donor states in dealing with international organizations. The arrangements for technical cooperation financing just described manage to avoid the budgetary scalpel because they are based on voluntary contributions. Nevertheless, indicative planning is still regarded by many developing states as an interim measure, which does not resolve the issue of long-term funding for technical assistance and cooperation.

Two matters that suggest the fragile nature of the current consensus deserve mention. One is the fact that not all members in a position to do so pledge support, and among those who do there has in recent years been a less-than-satisfactory record of pledging on a timely basis and of paying pledged contributions. The second relates to the footnote A projects introduced earlier. Footnote A projects, it will be recalled, are those that have been approved by the board for support but for which adequate funding is unavailable from the agency's technical cooperation fund. Funding of such projects is dependent on the willingness and ability of advanced states to pick up one or another project. In fact, a substantial proportion of footnote A projects are funded annually, most of them on a preferential basis defined by whether or not the recipient state is party to the NPT. This practice has been challenged systematically by countries such as India, Pakistan, and Argentina as discriminatory and a violation of the statute of the agency. Countries like the United States, which carry out the practice of providing footnote A assistance only to states meeting certain criteria, disagree, asserting that extra-budgetary resources can be deployed according to the wishes of the donor.[97]

In many respects, the footnote A procedure has served as an important safety valve, funding technically meritorious projects, facilitating the development of nuclear activities and programs in member states, and lowering the incentives of recipients to make a political issue of the level and method of technical cooperation funding. The number and value of projects in footnote A has roughly doubled between 1981 and 1986. On the one hand, this increase may reflect growing interest in nuclear development and greater sophistication in putting forward technically high-quality proposals. But it also indicates growing demand that, in times of financial stringency, may go unsatisfied. If that were to occur, it would not be long before frustrations built up and new tensions surfaced, especially if safeguards costs were to continue to rise, as they are destined to do as increasingly complex nuclear facilities come under agency safeguarding responsibility. Hence, the long-term issue of achieving "comparably predictable and assured resources" to what would exist if technical assistance were under the aegis of the regular budget remains to be resolved.[98]

Political and Symbolic Benefits of an Acceptable Balance

What might appear simply to be a numerical balancing game is, in fact, an important political and symbolic issue. The agency's membership is heterogeneous. Different groups of states espouse different values and seek different objectives. The advanced states emphasize safeguards, although not necessarily for identical reasons. The United States and the Soviet Union give the highest priority to IAEA safeguards because of their importance to the international nonproliferation regime, which they regard as essential to their global security interests. Most other major supplier states that support safeguards do so in recognition of the security benefits, but also because they are persuaded that international nuclear commerce, which is their central concern, requires an effective international safeguards system.

Developing country member states emphasize promotion, development, and assistance. So their support for safeguards has become the quid pro quo for assistance. In their view the reciprocal also is true—assistance is the quid pro quo for supporting safeguards. Importantly, most of these countries appreciate that technical assistance has increased as much as it has because of increases in support for safeguards, and that without safeguards the agency's technical assistance program would be only a fraction of what it is. But there are differences here also. Some members are NPT parties and others are not. The more technically advanced NPT states among them realize that their opportunity to capitalize on nuclear energy depends on their support for international safeguards. Some of them also are persuaded that effective safeguards are in their national security interest. That certainly appears to be the message of the third NPT review conference held in Geneva in August and September 1985, the final declaration of which underscored the conviction of the conference that "IAEA safeguards provide assurance that states are complying with their undertakings . . . promote further confidence among states and . . . help to strengthen their collective security," and that the acceptance of the IAEA safeguards on all peaceful nuclear activities in non-nuclear weapon states "is a major contribution by those states to regional and international security."[99]

Even in the face of this ringing affirmation, statements to the effect that safeguards are in everybody's interest are not as deeply embedded or secure as one might hope. Many states, even while endorsing safeguards, have given little thought to the security benefits they provide. And, in many cases, because they have modest or even negligible interests in nuclear energy, they are vulnerable to arguments—particularly from the non-NPT Third World states—that the priority given to safeguards by the IAEA distorts the purposes of its statutes. The argument continues that this priority biases resource allocations, depriving these states of opportunities that otherwise would be available to make effective use of nuclear technology. This kind of reasoning, even though largely self-serving on the part of its promoters, evokes emotional charges of neo-imperialism and encourages polarization along North-South lines. Controversy over resources and opportunity, as we have noted earlier, typically generates greater solidarity among the developing countries per se than the solidarity of those that are parties to the NPT feel toward their advanced-state NPT partners.

This situation defines the problem of balance in the agency today. It underscores the political need to establish and maintain an acceptable equity between the two statutory purposes of the IAEA. It also underscores the symbolic importance of maintaining substantial, and perhaps even increasing, levels of funding for technical assistance in order to more firmly lock in the support of Third World countries for a vigorous safeguards program. Freezing technical assistance, as some of the advanced states would prefer, would give the impression that technical assistance is deemed unimportant, and would reinforce the argument of those who contend that the only interest of the major powers is to promote their nonproliferation objectives through the agency, even at the cost of hampering its peaceful nuclear activities.

The Limitations of Technology Transfer

One of the great concerns in much of this is the practical reality that increased technical assistance for many countries will not bring them substantially closer to effective use of nuclear energy. They may lack the infrastructure to usefully absorb additional resources, or the capacity to identify and formulate programs and projects that could make constructive use of the assistance that already is available. (There is no shortage of complaints, from officials who are in a position to know, that there is less efficiency in the use of agency resources than one would like to admit.) Recognition of this is one of the reasons for the reintroduction, starting in 1985, of a training component for many agency technical assistance projects. In this situation, perhaps the most useful thing for the IAEA to do would be to intensify efforts already in place, as a consequence of a major board review of technical assistance in 1983, to provide special assistance in project identification and longer term planning for, and with, countries with limited nuclear experience.[100] Parenthetically, in examining infrastructure capabilities and requirements, the agency is in an excellent position to assess optimal energy-development strategies in many countries and to evaluate the relative usefulness of pursuing nuclear approaches or exploring, with appropriate expertise, non-nuclear approaches.

For Third World states that have established the requisite infrastructure and engineering components of a nuclear power program, there still remains the problem of financing of nuclear power programs. Agency recognition of this issue can be traced to 1969 when the general conference asked the director general to study "the likely capital . . . requirements for nuclear projects in developing countries during the next decade, and . . . ways and means to secure financing for such projects from international and other sources."[101] This question was raised again at the 1985 NPT Review Conference, the final declaration of which "recognizes . . . the difficulties which developing countries face . . . particularly with respect to financing their nuclear power programmes" and urges "the establishment of favourable conditions . . . for financing of . . . nuclear power programmes in developing countries."[102] Pursuant to a review conference recommendation, the IAEA has proposed to establish a special group of experts to study ways to assist developing countries in financing nuclear power programs.[103] While involvement in financial arrangements for implementing nuclear power projects lies beyond the agency's statutory responsibilities, the problem of dealing with the financial handicap question is one more area that may pose an important challenge for North-South relations in the agency, which could affect the overall question of balancing safeguards and technical assistance and cooperation.

NOTES

1. The South African incident demonstrates the extent of U.S.-Soviet cooperation in the IAEA. It was a Soviet satellite that first picked up the activity in the Kalahari desert and led to Soviet notification to the United States so that further investigation could be made. One could easily imagine that the Soviet Union could have sought to capitalize politically on the situation, but it instead chose to work toward the nonproliferation goal it shared with the United States.

2. The discussion that follows is based to a significant extent on more than forty interviews conducted by the author with present and former members of the IAEA secretariat, and with a cross-section of resident representatives and general conference participants in Vienna during the autumn of 1984 and the spring of 1985.

3. Cited in Arnold Kramish, *The Peaceful Atom and Foreign Policy* (New York, Harper & Row, 1963) p. 218.

4. See Glenn T. Seaborg, *Travels in the New World*, Report PUB-113 (Berkeley, Calif., Lawrence Berkeley Laboratory, 1977) p. 486.

5. IAEA, GC(XX)/RES/336 (1976). See chapter 3 for a discussion of the structure of the board and of the criteria governing representation thereon.

6. IAEA, GC(XXIII)/OR.211 (1979). The decision to reject the South African delegate's credentials was phrased in terms of his representing "a racist and illegal regime."

7. The vote was eighteen to eleven with five abstentions. See IAEA, GOV/DEC/113 (XXIV) September 1981.

8. IAEA, GC(XXIX)/RES/442 (1985).

9. In 1985, the United States abstained rather than vote against a resolution on South Africa (as it had done in 1981 in the case regarding South African participation in the IAEA's Committee on Assurance of Supply), perhaps because of its desire to marshal as much support as possible in dealing with the Israeli issue discussed below.

10. U.N. RES/487 (June 19, 1981).

11. The IAEA Statute does not make provision for expelling a member state, but does provide for suspension from rights and privileges of membership either for noncompliance with safeguards undertakings (IAEA Statute, Article XXII.C) or for persistent violation of the statute or of agreements the state has entered into pursuant to the statute (IAEA Statute, Article XIX.B). A member state also may be suspended from privileges if it is in arrears in payments of its financial contributions to the agency beyond a certain point (IAEA Statute, Article XIX.A).

12. These points are raised in IAEA, GC(XXV)/OR.237, paragraphs 23 and 54 (1981), and in IAEA, GC(XXVI)/OR.245, paragraph 8 (1982).

13. IAEA GC(XXV)/RES/381 (1981).

14. The vote was forty-three to twenty-seven with sixteen abstentions whereas forty-seven supporting votes were needed.

15. IAEA, GC(XXVI)/675 (1982).

16. IAEA, GC(XXVI)/RES/404 (1982).

17. The certification referred to arose out of congressional reaction to the denial of Israeli credentials. In approving the continuing appropriation for FY83, which included appropriations allocated to the IAEA, Senators Robert Kasten and James McClure attached an amendment prohibiting use of such funds for payment to the IAEA unless the board of governors certified to Congress that Israel is allowed to participate fully as a member nation in the activities of the agency. The director general's certification to the U.S. government was considered by the Congress to satisfy that requirement.

18. IAEA, GC(XXVII)/RES/409 (1983). The punitive measure was contained in operative paragraph 3. For reasons similar to those raised on the issue of suspension, the United States and other Western states regarded consideration of punitive measures in this case to lie outside the competence of the agency.

19. Complete texts of these statements were included in a letter from the Israeli resident representative to the IAEA to Director General Blix and circulated as IAEA document GC(XXVIII)/720 (Consequences of the Israeli Military Attack on the Iraqi Nuclear Research Reactor and the Standing Threat to Repeat this Attack for (a) The Development of Nuclear Energy for Peaceful Purposes and (b) The Role and Activities of the International

Atomic Energy Agency, Letter dated August 21, 1984, to the director general from the resident representative of Israel, August 30, 1984).

20. IAEA, GC(XXVIII)/RES/425 (1984).

21. The letter, dated September 24, 1985, was circulated to all delegates to the twenty-ninth general conference by the agency at the request of the Israeli resident representative. The document itself was not given a formal designation.

22. IAEA, GC(XXIX)/764. Consequences of the Israeli Military Attack on the Iraqi Nuclear Research Reactor and the Standing Threat to Repeat this Attack for: (a) The Development of Nuclear Energy for Peaceful Purposes; and (b) The Role of the International Atomic Energy Agency. Draft Resolution Submitted by Iraq. It is significant that the Israeli government's statement to the general conference explicitly asserted that "no State in the Middle East was excluded" from its policy to "not attack or threaten to attack any peaceful nuclear facilities in the Middle East or anywhere else." See IAEA, GC(XXIX)/OR.277, paragraph 34 (September 26, 1985).

23. IAEA, GC(XXIX)/RES/443 (October 9, 1985). (See also note 22 above.) Resolution Submitted Jointly by Denmark, Finland, Iceland, Norway, and Sweden, as revised.

24. Indeed, the issue surfaced once again at the 1986 general conference. In June 1986, Syria had indicated an intention to raise the Israeli question, this time not in the context of the attack on the Iraqi nuclear reactor, but in the context of Israel's failure to place all of its facilities under IAEA safeguards and the desirability of a nuclear-weapon-free zone in the Middle East. A draft resolution on this subject was prepared, calling on member states to discontinue cooperation with Israel on nuclear matters until that country complies with the provisions of the resolution. It was introduced at the thirtieth general conference in October 1986 (IAEA, GC(XXX)/792, The Israeli Nuclear Threat, a draft resolution submitted by Algeria, Iraq, Iran, Jordan, Kuwait, Lebanon, Madagascar, Morocco, Namibia, Saudi Arabia, Syria, Tunisia, and the United Arab Emirates, and subsequently joined by Quatar and Libya). The draft was opposed on the grounds that it singled out one member state of the agency and that it introduced extraneous political issues into the agency's activities. The United States moved, under rule 70 of the General Conference Rules of Procedure, that the issue at hand was not routine, but important, and therefore required a two-thirds majority for adoption. This motion was put to a vote and carried by a vote of forty-two to thirty-seven with eleven abstentions. The ultimate effect of this action was that the sponsors of the draft resolution asked for adjournment of the debate, thereby effectively removing the issue from the agenda for this general conference. This episode nevertheless underscores that while the Israeli issue now seems under control, it remains close to the surface and could erupt for any number of reasons at a future time.

25. Letter from the U.S. Assistant Secretary of State Gregory J. Newell, Bureau of International Organization Affairs, September 17, 1984.

26. For such a U.S. statement, see IAEA,GC(XXV)/OR.237, paragraph 51 (September 1981).

27. Ibid., paragraph 35, statement by Japan.

28. Ibid., paragraph 65, statement by Switzerland.

29. Ibid., paragraph 68, statement by Sweden.

30. IAEA Statute, Article VI.

31. Italy, in particular, pressed for an expansion that would ensure it a more permanent presence on the board and was the state that actually proposed the amendment.

32. IAEA, GC(XXIX)/752. Revision of Article VI As a Whole (August 1985).

33. At the thirtieth general conference, October 1986, the general conference adopted a resolution (GC(XXX)/RES/467) calling on the board of governors to establish an informal working group open to all member states to examine different proposals for revising Article VI of the statute and to present a report at the 1987 general conference. In the course of discussing this proposed resolution, several states (the United States, Soviet Union, and France among them) took the position that the present composition of the board effectively met the basic criterion of effectiveness.

34. IAEA, GC(XXV)/RES/386 (1981).

35. Interestingly, the only geographic region to experience a diminution in the number of professional staff between 1981 and 1985 is Eastern Europe.

36. For a graphic illustration of changes and trends in the four-year period, see "Staffing of the Agency's Secretariat," IAEA, GC(XXIX)/755 (September 16, 1984), especially Annex XIII.

37. Ibid. Annex XI. The area in which imbalance is perhaps most pronounced is the Department of Safeguards. Approximately three-quarters of the safeguards professional staff are drawn from advanced states of both East and West. Between September 1984 and September 1985, for example, twenty-two of thirty-five new appointments in that department went to advanced state nationals including *all* the fourteen most senior posts.

38. The emphasis on promotional and developmental activities would, of course, be fully consistent with the statutory objectives of the agency. The real question is whether the balance would shift to the detriment of safeguards.

39. Temporary blockage of Hungarian membership had to do with Western reaction to Soviet action in the case of Hungary's failed bid for independence in 1955. In striking contrast is the observation made by Atomic Energy Commission Chairman Glenn Seaborg in his report to President Johnson on the 1968 general conference of the IAEA. In that report, he noted that "the Czechoslovakian situation [that is, the Soviet invasion of Czechoslovakia earlier that year] had little impact upon the Conference itself since the IAEA

has now been accepted by most members as a technical organization and not a political one." Cited in Glenn T. Seaborg, *Travels in the New World* (see note 4 above) p. 771.

40. A number of these points are discussed in Lawrence Scheinman, "IAEA: Atomic Condominium?" in Robert W. Cox and Harold K. Jacobson, *Anatomy of Influence* (New Haven, Conn., Yale University Press, 1973) chapter 7. Most recently, only assistance to Cuba appears to be affected by the congressional mandate.

41. Warren Donnelly, "Nuclear Weapons Proliferation and the International Atomic Energy Agency: An Analytic Report," prepared for the United States Senate Committee on Governmental Operations (Washington, D.C., U.S. Government Printing Office, 1976) p. 92.

42. While even the harshest critics of safeguards do not write them off entirely, some occasionally come close. See, for example, the statement of Senator Gary Hart in Hearings on the IAEA Safeguards Program, United States Senate Committee on Foreign Relations, 97 Cong. 1 Sess. (December 2, 1981); and statement of Congressman Edward J. Markey in Hearings on Legislation to Amend the Nuclear Non-Proliferation Act of 1978 before the U.S. House Committee on Foreign Affairs, Subcommittee on International Security and Scientific Affairs and on International Economic Policy and Trade, 97 Cong. 2 sess. (August 3, 1982). See also Peter Pringle, "Nuclear Unsafeguards" *The New Republic*, December 23, 1982.

43. See discussion on the meaning of proliferation in chapter 6.

44. See statements of Director General Blix at the 1983 and 1984 IAEA General Conferences, in particular IAEA GC(XXVII)/OR.247 and IAEA GC(XXVIII)/OR.257.

45. This raises the further question of balancing transparency with protection of proprietary information, a topic discussed elsewhere in this book.

46. A counterargument is that the agency cannot know or predict the intent of a state and must, therefore, treat all states as potential adversaries. This creates a paradoxical situation, since effective safeguards also require cooperation of the safeguarded state. The solution to this paradox rests in state understanding that if safeguards are to work by the IAEA communicating to other countries the verified peaceful nature of the inspected country's nuclear programs, the latter country must facilitate agency activities, even if the activities have an adversarial character.

47. See E. V. Weinstock and J. M. deMontmollin, "IAEA Safeguards: Perceptions, Goals and Performance." Paper presented at a Brookhaven National Laboratory-Cornell University Workshop on Nonproliferation and Foreign Policy (April 1984). Quite aside from detection, the agency can provide reasonable assurance that materials in peaceful use are being so used that it can determine that a small amount of material unaccounted for is actually a diversion as opposed to a book loss or the result of faulty instrumentation and measurement procedures.

48. On timely warning, see Leonard Weiss, "The Concept of 'Timely Warning' in the Nuclear Non-Proliferation Act of 1978" (April 1, 1985). (Unpublished paper).

49. Treaty on the Non-Proliferation of Nuclear Weapons, Article III.1.

50. INFCIRC/153, The Structure and Content of Agreements Between the Agency and States Required in Connection with the Treaty on the Non-Proliferation of Nuclear Weapons, paragraph 28.

51. Typical values for a significant quantity range from eight kilograms of plutonium in separated form to twenty tons of thorium; detection time ranges from one month for separated plutonium to one year for natural uranium; and detection probability, which is the numerical parameter for risk of early detection, is 90–95 percent. A false alarm probability also is included. SAGSI had recommended seven to ten days detection time for separated plutonium. This was the United States' preferred position and it is tied to the time accepted as required to convert separated plutonium into the metallic component of a nuclear explosive device. The agency instead used criterion of "on the order of weeks" until it recently adopted a one-month detection time for separated plutonium. It should not be overlooked that reprocessing and plutonium separation are largely confined today to nuclear weapon states and a few others, mainly the Federal Republic of Germany and Japan. India also is engaged in plutonium separation, but, unlike the others, is not party to the NPT; nor does it have its activities under safeguards except when the material being reprocessed is itself subject to safeguards. The main problem on detection relates to large bulk handling facilities, whereas most of the facilities subject to safeguards are sufficiently small that the risk of detection is substantial. In other words, much of the debate relates to circumstances that may arise in the future but have not yet come into play.

52. IAEA, *IAEA Safeguards: Glossary* IAEA/SG/INF/1 (Vienna, IAEA, 1980) paragraph 12.

53. On these points, see Myron B. Kratzer, "New Trends in Safeguards," speech delivered to the Institute of Nuclear Materials Management, Albuquerque, N.M., July 1985; see also E. V. Weinstock and J. M. de Montmollin, "IAEA Safeguards: Perceptions, Goals and Performance" (see note 47 above).

54. Myron B. Kratzer, "New Trends in Safeguards," (see note 53 above).

55. Hans Grumm, "Basic Concepts of Safeguards," Fifth Annual Symposium on Safeguards and Nuclear Material Management, ESARDA (Versailles, April 1983) where it is reported that inspection goals were fully attained in 1978 in only 17 percent of the facilities inspected, and then rose to 45 percent in 1982. For 1984 this figure is 53 percent. Similar improvements at even higher levels have been achieved for sensitive nuclear materials at these facilities. Failure to meet inspection goals in no way implies a diversion, a point that is frequently misunderstood by the general public.

56. Ibid.

57. The safeguardability issue was raised primarily by Victor Gilinsky, a former member of the Nuclear Regulatory Commission. See, for example, his remarks on "Safeguarding the Fuel Cycle" before the Atomic Industrial Forum Conference on Nuclear Safeguards, Orlando, Florida (April 12, 1976). He notes that steps outside the classic system of safeguards may be required where reprocessing and fuel fabrication are concerned. See also his "International Discipline Over the Uses of Nuclear Energy," in Albert Wohlstetter and coauthors, eds., *Nuclear Policies: Fuel Without the Bomb* (Cambridge, Mass., Ballinger, 1978) pp. 73–77. The argument has been carried into the public debate most vigorously by Paul Leventhal, president of the Nuclear Control Institute. See, for example, his statement before the Subcommittee on International Security and Scientific Affairs and on International Economic Policy and Trade, U.S. Congress, House Committee on Foreign Affairs, "The International Atomic Energy Agency: Improving Safeguards," 97 Cong. 2 sess., March 3 and 18, 1982.

58. See, for example, Grumm (note 55 above).

59. For an earlier statement of concerns, see Leonard Weiss, "Nuclear Safeguards: A Congressional Perspective," *Bulletin of the Atomic Scientists* (March 1978) pp. 27–33; Victor Gilinsky (see note 57 above).

60. The following discussion attempts to synthesize criticisms made in congressional testimony, media accounts of the IAEA, and articles written on the subject. See, for example, testimony of Roger Richter, a former IAEA inspector, before the House Committee on Foreign Affairs, Subcommittees on International Security and Scientific Affairs, on Europe and the Middle East, and on International Economic Policy and Trade on June 17, 1981, following the Israeli airstrike on Iraq's nuclear reactor in June 1981 as reprinted in *Nuclear Safeguards: a Reader*. Report prepared by the Congressional Research Service, Library of Congress (Washington, D.C., U.S. Government Printing Office, December 1983) pp. 704–710; statement of Roger Richter before the Senate Committee on Foreign Relations, June 18, 1981, ibid., pp. 719–724; statement of Emanuel R. Morgan, also a former IAEA inspector, in hearings before the Senate Committee on Foreign Relations on IAEA programs of safeguards (December 2, 1981), ibid., pp. 748–752; Robert Ruby, "The Nuclear Reckoning," *The Baltimore Sun* (1981); Paul Leventhal, "Strengthening Safeguards: A High Priority for the NPT Review Conference," Special Report of the Nuclear Control Institute (1985).

61. In NPT agreements, access for routine inspections also is limited to designated strategic and key measurement points. Of course, if the situation goes beyond routine inspection and further information is required, this alters the statement on limitations.

62. Statement of Emanuel Morgan, former IAEA inspector, in a report prepared for the Nuclear Regulatory Commission, as reported in *Nuclear Engineering International* (May 1982) p. 15.

63. See Morgan, ibid.; see also statement of Roger Richter before the House of Representatives cited in note 60 above.

64. The discussion on the assessment of charges is drawn to a large extent from IAEA, *IAEA Safeguards 1980–1985: A Progress Report* STI/PUB/749 (Vienna, IAEA, 1986), and from discussions with members of the secretariat and other persons professionally involved in safeguards activities in the United States and Europe.

65. See Kratzer, "New Trends in Safeguards," note 53 above.

66. INFCIRC/153, paragraph 73(b).

67. The concern about efficiency is sufficiently great to have merited a separate statement in the Final Declaration of the Third Review Conference on the Treaty on the Non-Proliferation of Nuclear Weapons. Paragraph 10 of the Final Declaration relative to Article III and preambular paragraphs 4 and 5 of the NPT state: "The Conference commends IAEA on its implementation of safeguards pursuant to this Treaty and urges it to continue to ensure the maximum technical and cost-effectiveness and efficiency of its operations, while maintaining consistency with the economic and safe conduct of nuclear activities." Final Document, Review Conference of the Parties to the Treaty on the Non-Proliferation of Nuclear Weapons, Part I, NPT/CONF.III/64/1 (Geneva 1965) Annex I, p. 4.

68. Myron Kratzer, "New Trends in Safeguards" (see note 53 above).

69. See James de Montmollin, "Some Thoughts on Issues Involving IAEA Technical Support" (July 29, 1985) (Unpublished paper).

70. See references in notes 60 and 62 above.

71. See NPT/CONF.III/64/1 (note 65 above), Annex I, statements relative to Article III and preambular paragraphs 4 and 5 of the NPT, paragraphs 2 and 11 (pp. 3 and 4).

72. See Joint U.S.-Soviet statement that included specific reference to strengthening the IAEA and supporting implementation of international safeguards. *Times* (London) November 22, 1985.

73. IAEA, GC(XXVIII)/OR.257 (September 24, 1984) paragraphs 73 and 74.

74. See, for example, statement of the representative of Pakistan, IAEA, GC(XXVIII/OR.260, paragraph 47, and the statement of the representative of Brazil, ibid., paragraph 83.

75. INFCIRC/254 Appendix, *Guidelines for Nuclear Transfers* (February 1978). In truth, it is less the fact of publication than what is being published that irritates the critics. They do not want nonproliferation regime matters highlighted in the agency. But any state can ask the agency to publish anything related to nuclear matters, and under the circumstances the agency is bound to circulate it. Circulation does not mean endorsement.

76. "Revised Guiding Principles and General Operating Rules to Govern the Provisions of Technical Assistance by the Agency," Approved by the Board of Governors, February 21, 1979. IAEA document GEN/PUB/12/REV.2a, Principle A.1(i).

77. See, for example, INFCIRC/247, "The Text of the Safeguards Agreement of 10 February 1977 Between the Agency, Canada and Spain" (May 5, 1977); and INFCIRC/251, "The Text of the Agreement of 22 July 1977 Between Argentina and the Agency for the Application of Safeguards in Connection with a Cooperation Agreement Between Argentina and Canada" (November 25, 1977).

78. Paragraph 2 of the agreements cited in note 74.

79. IAEA, GC(XXIII)/OR.214 (1979) paragraph 41.

80. IAEA, GC(XXVI)/OR.240 (1982) paragraph 59; and IAEA GC(XXIII)/OR.215 (1979) paragraph 70.

81. There is, of course, nothing in the statute to prevent one state from assisting another bilaterally according to criteria it selects. The agency cannot discriminate, but it cannot require states to void all criteria in making their own contributions to particular states.

82. These figures were derived from IAEA, *Review of the Agency's Activities* IAEA, GC(XXVIII)/78 (July 1984) pp. 55–61. This review was undertaken pursuant to a request of the general conference in GC(XXVI)/RES/399 (1982). The activities covered include all of the agency's substantive programs in nuclear power and the fuel cycle; nuclear safety and waste management; radioisotope and radiation applications; the International Centre for Theoretical Physics; and safeguards. The discussion in this section focuses on 1957–1982 because of the convenience of being able to rely on the agency's own official cumulative record in the preceding document; the document reflects all of the adjustments necessary to account for changes in record-keeping and reporting. However, where appropriate, both in this section and elsewhere in the book, reference to the subsequent four-year period is made.

83. Contributors in 1985 included Australia, Canada, the Federal Republic of Germany, France, Italy, Japan, Sweden, Switzerland, USSR, UK, and the United States. The United States accounts for approximately two-thirds of what is contributed. In addition to extraordinary resources, states make contributions in kind, including cost-free experts and consultants, stipends for fellowships, and training courses.

84. IAEA, GC(I)/RES/5 (October, 1957). See also Scheinman, "IAEA: Atomic Condominium?" in Cox and Jacobson, *Anatomy of Influence* (see note 40 above) pp. 222–223.

85. Ibid. The Trieste Centre remains very active and is instrumental in continually upgrading the scientific competence of students and practitioners from developing states, although developing states were not the only ones to press for its creation. Italy, too, lobbied hard with support from Scandinavian states and, of course, Pakistan. Each had its own agenda: developing states and Scandinavians wanted to see more responsiveness to Third World interests; Pakistan in addition stood to have one of its nationals named director (Abdus Salam); and Italy wished to consolidate the legitimacy of its

control of Trieste. The Cairo Center, in contrast, was turned down by all of the advanced states at first and was a classical North-South issue.

86. IAEA, *Review of Agency Activities* (see note 82 above) pp. 109–112.

87. IAEA, GC(XI)/RES/230 (1967) Review of the Agency's Activities, Results of the Review of the Agency's Activities, paragraph 1.

88. See IAEA, *Review of Agency Activities* (see note 82 above) pp. 100–101.

89. IAEA, GC(XXV)/RES/386 (1981) Staffing of the Agency's Secretariat. Operative paragraph 2 "reaffirms that no post, including that of the director general, should be considered the exclusive preserve of any member state or group of states, and that all posts at all levels should be considered open to appropriately qualified candidates from all member states."

90. See, for example, statements by representatives from Malaysia, IAEA, GC(XXIII)/OR.215, paragraphs 58 and 59; Nigeria, IAEA, GC(XXIII)/OR.216, paragraph 143; and the Philippines, IAEA, GC(XXIII)/OR.221, paragraph 95. All three are parties to the NPT in different stages of nuclear development. See also India, IAEA, GC(XXIII)/OR.220, paragraph 24.

91. See statement of representative of Pakistan at the twenty-eighth session of the general conference, IAEA, GC(XXVIII)/OR.260, paragraph 50 (September 26, 1984).

92. As mentioned earlier, several European states, led by Belgium, have urged consideration of an approach that would place the burden of real-cost increases on the shoulders of the nuclear weapon states. This appears to be motivated by a number of factors, among them the sense of some advanced industrial states that they are being overinspected and are overpaying for the privilege; and is defended by them on the grounds that the safeguards accepted by the weapon states on their civilian nuclear facilities are not comprehensive and therefore are largely symbolic, and that until comprehensive safeguards are applied on the civilian facilities of *all* countries, a financing formula that differentiates between the weapon states and others is appropriate. Needless to say, the weapons states do not share this view.

93. See statements by representatives from Malaysia, Nigeria, the Philippines, and India cited in note 90 above.

94. IAEA, GC(XXV)/RES/388, The Financing of Technical Assistance (September 26, 1981).

95. The notion of regular budget funding of technical assistance once was broached by the United Kingdom primarily to rope the Soviet Union into providing a fair share of the assistance.

96. See, for example, the statements of representatives from Pakistan and Nigeria cited in note 90.

97. The U.S. position is the correct one. Preferential treatment is prohibited under the statute with respect to the dissemination of agency resources. Extrabudgetary contributions, however, can be designated for specific countries according to whatever criteria the donor chooses. Even before the NPT

existed, donor states made specific equipment available to particular recipient states through the IAEA.

98. The twenty-ninth general conference voted a resolution requesting the board of governors to report annually to the general conference on actions taken to implement RES/388 of 1981. IAEA, GC(XXIX)/RES/452, The Financing of Technical Assistance (October 11, 1985).

99. See Final Declaration of the Third Review Conference of the Treaty on the Non-Proliferation of Nuclear Weapons (see note 67 above) (Article III and preambular paragraphs 4 and 5, paragraphs 2 and 3, respectively) NPT/CONF.III/64/I, p. 3.

100. An indicator of the success rate of this program may be the increased number of technically meritorious projects mentioned earlier in connection with discussion of footnote A projects.

101. IAEA, GC(XIII)/RES/256, The Agency's Budget for 1970, The Financing of Nuclear Projects (October 6, 1969).

102. See Final Declaration of the Third Review Conference of the Treaty on the Non-Proliferation of Nuclear Weapons (see note 67 above) (Article IV, paragraph 21) NPT/CONF.III/64/I, p. 9.

103. IAEA, GOV/INF/487 establishing a Senior Expert Group on Mechanisms to Assist Developing Countries in the Promotion and Financing of Their Nuclear Power Programs (January 27, 1986).

Chapter 8

RETROSPECT AND PROSPECT

In reflecting on the evolution of the International Atomic Energy Agency (IAEA), two aspects stand out: the changes that have occurred in the agency and in its political environment over the years, and the increasing relevance of safeguards and of the controversy that surrounds them.

POLITICAL EVOLUTION

As for change, there are three discernible periods in the IAEA's history.[1] The earlier years were dominated by the United States, whose initiatives largely were responsible for the creation of the agency. This was followed by a period of close Soviet-American cooperation and near hegemony in the agency in which several programs, particularly safeguards, were dominant. This new climate resulted in the establishment of the safeguards system that now applies to some extent in virtually every nation that has significant nuclear activity. The third period dates from the mid- to late 1970s, when a sizable group of Third World member states began to make their presence felt in the agency.

The Soviet Union's 1963 decision to support IAEA safeguards signalled the major changes that took place in the second period. A constructive shift, it had an impact on the general mode of IAEA's operation, including strong support of decision making by consensus rather than by formal voting procedures. It also established the basis for a close relationship between the United States and the Soviet Union regarding both safeguards and broader nonproliferation issues—a relationship that has been exceptionally stable despite the tensions that exist elsewhere between the two superpowers. It survived the Soviet invasion of Czechoslovakia in 1968 (as did détente) as well as President Carter's decision to withdraw SALT II from Senate consideration for several reasons including the 1979 Soviet invasion of Afghanistan, and the deterioration of East-West relations symbolized by President Reagan's "evil empire" speech, his refusal to submit SALT II to the Senate for its advice and consent, and the U.S. arms control posture in general. In view of the importance of avoiding proliferation to both national security and general international stability, the U.S.-Soviet concurrence of outlook and cooperation cannot be emphasized too strongly.

The current period began under unfavorable circumstances and has been marked by more traumatic events. With the near universal spread of decolonization and the principle of self-determination, the political and economic structure of the world has been fundamentally altered. The family of nations has grown, and new spokesmen with new agendas have emerged, demanding a restructuring of the international economy, as well as the vigorous pursuit of development and disarmament programs. Reflecting a solidarity based on their resentment of perceived dependence on the North, and their former colonial experience, the agenda shared by nations of the South is found in programmatic claims and institutional developments.[2]

The springboard for the Third World was the first U.N. Conference on Trade and Development (UNCTAD), which was held in Geneva in 1964, and which established an extensive agenda for global economic reform that was elaborated at later UNCTAD meetings.[3] The Group of 77, whose membership now exceeds 100, was formed in 1964 to further the promotion of the interests of the developing world by establishing a unified bargaining front. Such development paved the way for a surge of activity in the 1970s, including the Action Program for a New International Economic Order adopted at the 1973 Summit of Neutral and Non-Aligned Countries in Algiers, the Charter of Economic Rights and Duties of States adopted by the U.N. General Assembly in 1974, and the Conference on International Economic Cooperation, held in Paris from 1975 to 1977.

While a number of agencies in the U.N. family are deeply concerned with development and have served the interests of the Third World, the past decade has witnessed the growing use of the practice of majoritarianism[4] in the assemblies of the world's international agencies and organizations to further the agenda of developing countries. Through these means, the Third World has poignantly challenged the regimes developed by the Western world (largely at U.S. initiative) to preserve, institutionalize, and strengthen established values, the prevailing order, and the status quo.

The nuclear arena is part of this landscape. By the mid-1970s, there was increased Third World interest in nuclear energy and, in some cases, in access to the full range of nuclear technologies, including those that could directly involve weapons-usable materials. At the same time, in the United States and elsewhere, there was increased concern about the adequacy of the nonproliferation regime, with its reliance on commitments and safeguards, to deal with the spread of technologies that could provide a basis for nuclear weapons capability. This convergence of opinion culminated in an agreement among the major supplier states to require as a condition of nuclear transfer a "no nuclear explosives" pledge—thus eliminating the possibility of pursuing a peaceful nuclear explosives program—and perpetual safeguards on all transferred material or facilities. In addition, the suppliers also agreed to exercise restraint on a voluntary, not mandatory, basis, in the transfer of sensitive technologies (reprocessing, enrichment, heavy water production) and facilities.[5]

Third World countries—feeling alienated and growing more assertive—viewed the suppliers' meetings and results as a cartel-like action. They felt that such agreements discriminated between industrial and developing countries and was an extension of more general economic discrimination into the peaceful nuclear realm.[6] Third World states who were signatories to the NPT also viewed this action as a violation of the NPT's commitment to peaceful nuclear development.

Thus, the matter was transformed into a North-South issue that has affected relationships and behavior in the agency ever since, principally by reinforcing tendencies toward polarization that have characterized North-South relations in general. It provided the Third World countries—which in the IAEA are normally divided along NPT/non-NPT lines and according to their degree of nuclear advancement—with the glue that would bind them together and facilitate bloc-type voting on a number of issues. The first formal caucus of the Group of 77 in the IAEA context was held in 1976;[7] and the first formal mention of the Group of 77 before the board occurred in 1978. In 1981, when the election of a new director general was at stake, polarization peaked. The election, for which a serious Third World candidacy was mounted, took place in the aftermath of the Israeli bombing of the Iraqi research reactor at Tamuz. These two events combined to bring about the most severe confrontation that had yet been experienced within the agency between developed and developing-member states. The fight over the director general's post also tied polarization and politicization together. While some of the issues have been temporarily resolved (for example, improved levels and predictability of technical assistance through indicative planning, and the increased presence of Third World nationals in high-level secretariat posts), the basic situation has not changed, and potential bloc voting based on bloc politics is still a factor. Particularly troublesome is the fact that this caucus has held together to keep such extraneous political issues as South Africa's apartheid policy and Israel's position in the organization on the agenda. An important step away from this pattern and toward restoring some priority to the agency's statutory objectives was achieved with the adoption of the Nordic Resolution on the Israeli issue at the 1985 general conference.[8] In theory the corner has been turned, but only time will tell whether practice will conform to theory.[9] As will be argued below, a net positive outcome will require continuous constructive effort on the part of those states whose interests rest in a strong, effective, and reliable international nuclear organization. An important element in this challenge is ensuring that those interests are broadly based and include key states from each of the major constituencies that comprise the agency.

SAFEGUARDS

As world nuclear development and transactions have increased, safeguards have taken on greater relevance. And with their increased relevancy, they have become more controversial. Critics have questioned their effectiveness, their implementation, and their cost. Until the 1970s, the role of safeguards, though important, was set against the background of relatively modest nuclear development occurring largely in industrial states. Few were regarded as serious proliferation risks; indeed, most were linked to one of the superpowers in a security alliance. And at the time, the NPT was seen as substantially reinforcing the growing presumption against the legitimacy of weapons proliferation. Safeguards were viewed as an effective and useful means of verifying and confirming the peaceful nature of nuclear activities and of sounding the alarm in the event of a diversion.[10]

Subsequently, the wider dissemination of nuclear materials and technology and the trend toward fuel-cycle activities involving potentially widespread commerce in weapons-usable material—particularly plutonium—raised the question of safeguards effectiveness. The Indian nuclear explosion demonstrated to everyone—including other Third World states—that mastery of this awesome technology was not beyond their reach if they were determined to proceed. It also opened to question the adequacy of a regime that relied so heavily on safeguards implemented by an international organization with limited authority and no direct enforcement powers. Although no explicit violation of safeguards was involved in the Indian case, many thought of safeguards, and the nonproliferation regime of which they were a part, as interchangeable concepts. Consequently, they assessed safeguards effectiveness in terms of the overall purpose of the regime. Since the regime was designed to prevent proliferation, some reasoned that if safeguards should prove incapable of achieving that goal, then they should be deemed inadequate and undeserving of credibility.

Others who did not go quite so far as equating safeguards with the entire regime nevertheless questioned whether safeguards could effectively detect diversion in all circumstances. They concluded that if safeguards could not detect, with a high degree of certainty, a diversion of a significant quantity of weapons-usable material in a bulk-handling facility, then the activity involving the material itself should be terminated, or at least severely curtailed until such time as effective nonproliferation measures were devised and deployed.[11]

Yet another controversy over safeguards relates to the emphasis given to support for them in comparison with support for other IAEA activities. As we have seen, developing countries are adamant in their view that the IAEA should carry out a balanced program so that comparable attention is given to development of nuclear technology and peaceful activity. Their concern relates not only to whether there is comparable emphasis now, but what the case will be over the longer term, and especially the degree of confidence they can have in a stable and predictable flow of resources. Progressively costly safeguards, in their view, eventually will have an adverse impact on other agency programs. Therefore, those programs must be afforded ample protection and safeguards funding must be of such a level and nature that evenhanded attention to and support for the other principal areas of agency responsibility will not be threatened.[12]

Those who are committed to effective safeguards and effective nonproliferation face several substantial challenges:

- Defining the proper role for safeguards and achieving consensus on criteria for evaluating their effectiveness;
- Improving the capacity of safeguards to detect diversions of significant quantities of nuclear material at all types of facilities with a sufficiently high degree of probability to deter would-be diverters;
- Creating institutional arrangements to compensate for any residual weaknesses in technical safeguards; and

- Establishing financial mechanisms for safeguards that will ameliorate continuing concerns over runaway costs.

In virtually all quarters, it is beyond question that safeguards are an essential factor of international nuclear cooperation and nonproliferation. Without them, there could have been relatively little international nuclear cooperation, and such nuclear development as did occur would have continued unchecked. The spread of nuclear weapons likely would have been greater than that which has occurred; and even more certainly, mutual suspicions about the nature and purpose of nuclear programs in neighboring states would have resulted in increased regional tension and destabilization. Seen in this light, the IAEA has made a major contribution to facilitating access to the peaceful uses of nuclear technology. And at the same time, it has limited the risk of increased proliferation by implementing a system of objective, impartial verification safeguards. This is no mean achievement, and it is one that we cannot afford to lose.

PROSPECTUS FOR THE FUTURE

How can the IAEA be strengthened? Toward this end, initiatives in four areas—leadership, safeguards, depoliticization, and the balance between safeguards and technical assistance—can be identified.

Leadership

Proposition. The long-term viability of the IAEA as an effective instrument for ensuring the safe and peaceful use of nuclear energy depends upon dedicated and sustained leadership by important member states. The United States, by virtue of its historic role in the agency, the scope and extent of its contributions, and its unique combination of scientific, technological, and industrial capabilities in the nuclear arena bears a special leadership responsibility. But that responsibility can be exercised effectively only under two conditions: (1) that as a nation and through its executive and legislative branches of government, it affirm its commitment to the IAEA in word and in action; and (2) that it recognize that neither unilateral nor bilateral superpower leadership will alone suffice any longer, and that a collective leadership reflecting the full cross-section of legitimate interests represented in the agency's membership is essential to future success. Emphasis must be placed on establishing a strong and robust consensus that draws substantially, and early in the decision-making process, on the major groups of agency membership, East and West, North and South.

Argument. Neither the politicization nor the safeguards credibility problems of the agency, nor any other problem for that matter, will be resolved unless there is a strong determination to do so. Objectives need to be clarified, priorities established, and commitment to effective and constructive participation reaffirmed. This, in turn, requires a vigorous assertion of leadership. Where is that leadership to come from? To a point, it can be, and in fact is, provided by the director general and the secretariat of the IAEA. But the IAEA, like all international organizations, is a creature of its constituents, the member states, and lacks the autonomous political base so essential to bringing the necessary resources to bear to carry initiatives through to successful conclusion. It can cajole, argue, influence, persuade, but it cannot decide authoritatively, other than where empowered to do so by its constituents. In a real sense, it is as strong or as weak as its membership wishes it to be, and sovereign states do not transfer authority lightly.[13] Thus, whatever the secretariat of the agency may do, leadership also must come from the member states. In the case of the IAEA, although there are recent indications that the sources of initiative may be broadening, that leadership responsibility devolves primarily on the United States.

Why, it might reasonably be asked, should this responsibility fall to the United States? After all, the United States has lost its once hegemonic role in the nuclear arena. Its share of the nuclear marketplace is considerably reduced; other nations are moving with more determination toward the development of advanced reactor concepts (such as the breeder); there is increased political ambivalence in the United States over the future of nuclear power in general; and U.S. interest and confidence in the IAEA appear to have diminished from their once enthusiastic level, reflecting a lack of confidence in existing international controls and an uncertainty in some quarters of the U.S. political establishment about the optimal strategy for controlling nuclear proliferation. If uncertainties are less evident in the current administration than in its predecessor, nagging doubts still linger and periodically flare up in the Congress.

The answer very simply is that no other state or group of states is likely to take and exercise that responsibility. As we have pointed out before, the history of international nuclear development and of nonproliferation is largely a history of U.S. efforts and initiatives. The Baruch Plan, Atoms for Peace, the IAEA, international safeguards, the NPT, the Nuclear Supplier Guidelines, INFCE, all emanated from Washington. Notwithstanding the difficulties and changes just noted, the United States is still better positioned technologically and politically than is any other country to assert leadership responsibility. If diminishment of the U.S. position in the nuclear field has created something of a void, nothing has come to fill it and nothing appears on the horizon. More importantly, that is the perception of a vast majority of the active international nuclear community. Others continue to look to the United States, with its extensive technological and industrial base and its national laboratories, for guidance, benchmarks, clues, and direction in their own nuclear development. With all the attendant difficulties, the United States is still widely viewed as being at the leading edge of nuclear technology and is a preferred nuclear partner.

Within the IAEA, the United States remains the principal donor, a vital force accounting for slightly more than 25 percent of the agency's resources and in addition providing cost-free experts, in-kind assistance, and research and development support in the critical area of nuclear safeguards. It was generally acknowledged in the wake of the U.S. walkout in 1982 (following the general conference's rejection of Israeli credentials) that without the U.S. presence in the IAEA, a void would be created that could not be filled, and the agency's days would be numbered. The United States should not take this as a sign that it can hold the agency hostage to its wishes, as it sought to do in 1984 and 1985 when it informed the director general that it was a firm and non-negotiable U.S. policy that any limitation on the rights and privileges of Israeli membership in the agency would cause the United States to suspend its participation and support. Rather, it should regard its central role as an opportunity to promote constructive activities and measures leading to a strengthened and even more effective International Atomic Energy Agency. Indeed, this will be essential if the IAEA is to remain a strong pillar of the nonproliferation regime. It is difficult to imagine how U.S. interests could be served by anything less than an active and constructive leadership role.

Leadership entails taking initiatives such as the United States proposed in 1981 (unfortunately without success at the time, but later vindicated following the Chernobyl accident as discussed below) for an international convention on nuclear safety cooperation and mutual emergency assistance in connection with nuclear accidents. But, as just suggested, it also means avoiding the exacerbation of tendencies such as politicization to which the United States contributed by seeking to foist domestic political choices on the agency with the threat of withdrawal if its will was thwarted.

Leadership means anticipating problems, not just responding to crises. The principal challenge to leadership today is to ensure that the IAEA is fully capable of keeping pace with an ever-growing workload and meeting its responsibilities, and that expectations and capabilities are kept in some reasonable balance. To do this requires resources. For the past several years, the agency has been on a zero-growth budget, although its allocations to safeguards have increased if only ever so slightly. The agency always has been tightly run in fiscal terms, and the consequence of zero-growth budgeting has been a progressive squeeze of technical programs that interest a broad cross-section of agency membership but that in many cases are of special concern to developing nations. Their support for safeguards remains essential but would be increasingly difficult to sustain if technical cooperation programs were to begin to atrophy.

In the wake of the Chernobyl accident, the agency has become an early and major focal point of international safety concerns and new demands inevitably will be placed on it.[14] In this regard, the exceptional initiative taken by General Secretary Mikhail Gorbachev shortly after the event is both striking and significant. In an unusual television speech on May 14, 1986, Gorbachev publicly urged a "serious deepening of cooperation in the framework of the IAEA," asserting that "it would be expedient to enhance the role and possibilities of that unique international organization. The Soviet Union is ready for this."[15]

Expanding on this theme several weeks later in a message to President Reagan, U.N. Secretary General Javier Perez de Cuellar and IAEA Director General Hans Blix, Gorbachev urged that the international community establish a convention to define the legal obligations of states regarding nuclear safety; proposed the development of an international regime for safer nuclear energy using such existing international institutions as the World Health Organization and the United Nations Environmental Program; suggested the development of recommendations for the security of nuclear power plants and for international control (surveillance?) for their application to all states; and called for unification of permissable levels of radiation in different countries, with the IAEA as the centerpiece for these efforts and activities, concluding that its "financial and material possibilities should be expanded." Taken at face value, this set of proposals must rank among the most important in more than a decade of agency activity.[16] Their presentation offers the world an important opportunity to advance a collective interest, and the United States, an earlier advocate of a number of these concepts, the occasion to play a major role in advancing the common cause. Even more importantly, it creates an opportunity for Soviet-American cooperation in the IAEA to reach beyond safeguards into new areas of common interest, thereby further strengthening the foundations of the agency and enhancing its ability to contribute to international welfare.

Nuclear safety is not new to the agency agenda. In the past, the IAEA has formulated basic safety standards for radiation protection and produced codes of practice related to a wide variety of different phases of nuclear activity. What has changed is the magnitude and perceived urgency of the issue of nuclear safety. As critical to international well-being as this issue is, it is important that the response of member states in the form of augmented resources for safety-related activities not result in failure to provide needed additional support for other agency responsibilities such as safeguards. The zero-growth budget mentality that has dominated agency financial deliberations (and programmatic development) during the past several years could conceivably lead to such a result. In this respect, it is well to recall that the NPT Review Conference, while commending the IAEA for its safeguards accomplishments, emphasized the importance of continued improvements in the effectiveness of IAEA safeguards; supported the notion of extending agency safeguards to additional civil facilities in nuclear weapon states beyond those now covered by voluntary offers; and underscored the need for the IAEA to be provided with "the necessary financial and human resources to ensure that the Agency is able to continue to meet effectively its safeguards responsibilities."[17]

The agency could be severely damaged if its tasks and responsibilities were increased without commensurate financial and professional manpower resources and support. That could only lead to overexpectation and perceived underachievement, a deadly combination as we have seen in the field of safeguards. The time has come to move boldly forward with initiatives to augment the IAEA's resources and to provide it with the level of support commensurate with the national interest, with international stability, and with the security functions that it performs—in short, to recognize that the IAEA is part of our national security and to treat it accordingly. This is a responsibility to which U.S. leadership must rise.

Granted the importance of leadership and the central role for the United States in that context, it is essential to understand that the international political environment has altered the circumstances and conditions under which leadership must operate. This cannot be emphasized too strongly. The United States is less able to dominate the agency than it once could, although it remains its single most influential member. Today, however, the United States requires support from other quarters if the objectives it espouses are to be achieved and if the agency's long-term viability is to be assured.

Consensus-building is more important now than ever. On the one hand, this means it is necessary to capitalize on the fortunate convergence of Soviet-American nonproliferation goals and newly stated Soviet interest in actively augmenting the general role of the IAEA in the peaceful nuclear arena as well as on the continued support of traditional U.S. allies.[18] But on the other hand, it means engaging the active involvement and support of moderate developing nations that share the general objective of nonproliferation, especially those that have serious interest in the peaceful uses of nuclear energy over the longer term and who understand that their attainment of that goal is the reciprocal of their support of effective nonproliferation safeguards without which international nuclear cooperation would most likely atrophy. These nations want to be major players in the policy-making enterprise, and part of the definition of leadership is to give active encouragement to participation and to listen to what other countries want and believe. Leaders in the United States nuclear and nonproliferation communities must adjust their own expectations and strategies in light of changing political circumstances. They must be prepared to invest substantial political capital to maintain the agency as a major contributor to nonproliferation and international stability. It is part of our national security. Self-reaffirmation of this conviction and these goals may be a first priority for the United States today.

Safeguards

Proposition. Effective international safeguards contribute importantly to national security and to international stability and therefore deserve whatever political, financial, technical, and human resource support is necessary to the fulfillment of their task. There is a need to clarify the role of safeguards in nonproliferation, and to clearly define and agree to the criteria by which to gauge their effectiveness. Efforts to increase safeguards transparency should be pursued in the interest of improving their credibility, which is essential to sustaining and building requisite public confidence. The search for, and evaluation of, supplementary institutions that can assume responsibilities that lie beyond the purpose and capabilities of safeguards, but that can enhance their effectiveness, should be continuously pursued. Above all, state cooperation in the implementation of safeguards must be brought to the highest possible level.

Argument. International safeguards have an invaluable and irreplaceable function in the broad fabric of nonproliferation. In the United States, the agency and its safeguards system have twice been the subject of major reviews and have been found necessary to U.S. national security interests. The most recent occasion was in 1982 following the U.S. walkout from the agency's general conference in the context of the dispute over treatment of Israeli credentials. The conclusion reached in that review by the Reagan administration was that if the agency did not exist it would have to be created, and that there was no viable alternative to fulfill its mission—in particular, its safeguards responsibilities, which were critical to international security and nuclear commerce. This conclusion was virtually identical with one reached a decade earlier by a commission impaneled to assess the efficacy of the IAEA for the achievement of U.S. policy objectives. That review also found the activities of the agency of vital importance to the security and well-being of all nations and underscored the urgency of vigorous U.S. support for the agency, "especially because of its critical role in carrying out the provisions of the Treaty on the Non-Proliferation of Nuclear Weapons."[19] The strong endorsement of agency safeguards at the 1985 NPT Review Conference demonstrates that these conclusions are widely shared in the international community. Indeed, the record of the past decade reveals that most other countries had greater confidence in safeguards (or somewhat different expectations) than did the United States, and were therefore even more prepared to rely on them as nonproliferation measures.

With respect to IAEA safeguards, steps need to be taken on three levels: (1) understanding the role of safeguards; (2) making intrinsic improvements to safeguards; and (3) exploring the potential contribution of supplementary institutions to enhanced safeguards.

Role of Safeguards

A first priority is to clarify the role and purpose of safeguards and of related detection and inspection goals. To communicate with each other, nations must first stop talking past each other. Clarification is as necessary, perhaps even more necessary, in the United States than in the international community as a whole. At the level of public understanding, a point of departure would be to ensure awareness that the role of safeguards is not to control (a term that invites images of prevention), but to contribute to confidence by verifying a state's accounting of the use of its nuclear materials, thereby providing objective third party assurance that undertakings and commitments to peaceful use are being sustained. This understanding is essential to the establishment of a basis upon which to judge safeguards effectiveness and to determine what steps are required to strengthen them and enhance their credibility.

A second step is to work toward establishing agreed criteria for determining safeguards effectiveness. This involves the proper understanding and use of safeguards goals. We may recall that with the advent of the NPT, important non-nuclear weapon member states of the agency urged that the safeguards system be reviewed in the interest of ensuring as objective and scientific a system as possible, reflecting their desire to protect against the risk of arbitrary or unnecessarily intrusive safeguards now that all of their nuclear activities would be covered. In contrast to INFCIRC/66, the safeguards document that was developed to meet NPT obligations provided a technical definition of the safeguards objective, namely "the timely detection of the diversion of significant quantities of nuclear material from peaceful nuclear activities. . . ."[20] The key terms of this objective were not defined, although such definition was essential to making the technical objective operationally useful and to achieving its purpose.

The task of definition was given to the Standing Advisory Group on Safeguards Implementation (SAGSI), an advisory body of technical experts nominated by their governments. SAGSI identified four terms requiring quantitative expression (timely detection, significant quantities, risk of detection, and probability of raising a false alarm), and defined the associated numerical parameters (detection time, significant quantity, detection probability, and false alarm probability) as detection goals. In 1977 the group submitted provisional numerical estimates for these goals to the agency.

These detection goals were tentatively and informally used (but not adopted) by the agency as guides for safeguards planning and as a yardstick against which it could measure progress in safeguards development. They also were used as a point of departure for establishing *inspection goals*, but these were based on actual conditions at facilities to be inspected and on existing safeguards measures. They were decidedly not intended as criteria of safeguards effectiveness. These nuances escaped many outside the small safeguards community, and the detection goals, which had received a certain amount of publicity, came to be viewed as the basis for judging safeguards effectiveness even though the agency had not actually established effectiveness criteria and was using criteria loosely reflecting the generally more liberal inspection goals.[21]

When differences between the perceived (that is, detection) goals and the actual results of safeguards exercises began to show up (as reflected in the secretariat's Safeguards Implementation Reports to the board, the agency's Annual Report, congressional hearings on nonproliferation, and public interest group tracts), earlier convictions about the inadequacy of safeguards, particularly in certain fuel-cycle areas, were reinforced. More information about inspection and associated criteria was made available, but it did not substantially alter earlier conceptions, and when it turned out that neither detection nor inspection goals were being fully met, critics were further persuaded that safeguards were inadequate and their concern was reinforced that undetected diversions could have taken place. Much of the controversy surrounding safeguards today flows from tendencies to misconstrue the purpose and meaning of safeguards goals, and from continued lack of agreed criteria for determining safeguards effectiveness.

Some proponents of safeguards consider the values assigned to safeguards goals as unduly restrictive and urge their liberalization, while others, equally supportive of safeguards, feel that altering the numerical values holds little promise for improving safeguards credibility. In the latter view, attaining consensus on new values would be difficult. Marginal changes would still leave a gap between what is preferred and what can be accomplished, while changes of a magnitude sufficient to cover such troublesome problems as large bulk-handling facilities (for example, facilities in which current safeguards capabilities only could detect at the level of three or four significant quantities) would risk reducing overall confidence in the credibility of safeguards. They instead advocate maintaining the present goals but educating the relevant public on the actual purpose of these goals and the place of safeguards in the broader nonproliferation regime. In our view, this is the preferable approach to the problem.

Accepting the values currently assigned to detection goals does not mean, however, that no adjustments are in order. For example, it has been suggested that in implementing safeguards more attention should be given to the scope and character of the fuel cycle in the inspected state than is currently the case.[22] The agency practice is to inspect similar facilities in different states in an identical manner. A main reason for this is the principle of nondiscrimination, a standard to which international organizations traditionally are closely held. But in the case of the IAEA, both safeguards documents provide for taking account of the fuel cycle in determining the frequency or intensity of inspection. This appears to offer the possibility to differentiate the application of safeguards in different states on the basis of objective considerations rather than on subjective political judgments about a state's nonproliferation reliability, and to place important boundaries around the argument that differential treatment per se constitutes discrimination.

Several arguments favor pursuing this conceptual approach. One is that it might help to control the problem of rising safeguards costs by focusing safeguards at the most vulnerable points in the national nuclear fuel cycle. Proponents of the fuel-cycle safeguards approach contend that the application of a facility-specific uniform approach that disregards the character of the fuel cycle in the inspected state could result in virtually all safeguards resources being expended in only a few countries. One advocate, for example, claims that full application of a uniform approach on a facility-specific basis in 1983 would have resulted in all of the agency's resources being used in three countries—Canada, Japan, and the Federal Republic of Germany—because of the large number of reactors in these non-nuclear weapon states.[23]

The resource allocation arguments take on added significance when one considers that the Final Declaration of the 1985 NPT Review Conference urged "further evaluation of the economic and practical possibility of extending safeguards to additional civil facilities in nuclear weapon states."[24] The nonproliferation value of such a measure may be open to question, but the main point is that the effect of such an action would be to further increase the resource burden that safeguards entail. Alternative safeguards deployment strategies that offer greater efficiency without sacrificing effectiveness thus become both timely and pertinent.

Of course, in considering the merits of fuel-cycle safeguards, one should not lose sight of the fact that it has not been possible up to now to achieve agreement on objective ways in which to take the fuel cycle into account. Nor has there been concurrence on what technical conclusions could be drawn by the randomly conducted inspections that such an approach would likely involve. Since the emphasis has been on objective, technically anchored safeguards, this is not something that can easily be set aside. It is also worth noting that even if there were agreement, it is still possible that the fuel-cycle safeguards approach could result in even more intensive safeguards being applied in full fuel-cycle states than results from the current system. This would be quite a different outcome from what many of its proponents anticipate, namely that an integrated approach would reduce the safeguards burden in states with large nuclear programs including substantial fuel-cycle facilities.

The second argument in support of fuel-cycle safeguards is that they would provide the agency with added flexibility, thereby increasing the degree of assurance that safeguards provide and thus reinforcing perceived safeguards effectiveness and credibility. This contention is based on the proposition that the deterrent effect of safeguards is increased because all possible diversion pathways would be theoretically covered while the agency would retain full control over the selection of the facilities and pathways that actually would be subject to a given inspection. If operators knew that there was a significant risk that diversion could be detected in every plausible pathway but were uncertain as to which pathway might be inspected, safeguards credibility would be enhanced in the eyes of third parties who depend upon safeguards to assure them that no material was being diverted.

An important underlying assumption of this argument is that it would be possible to draw technical safeguards conclusions when using short-notice inspections at randomly selected facilities, and this is still open to question. In addition, while flexibility undoubtedly would be increased, it is possible that the risk of detection might decrease. These considerations raise important policy questions about what our priorities ought to be.

Whether or not the fuel-cycle safeguards approach would result in a significant improvement in safeguards credibility as its proponents contend, is not something that can or need be decided here. The only point is that concepts such as this deserve close scrutiny and evaluation, and the generation of other ideas should be positively encouraged. Among the other proposals that merit exploration is the notion of making greater use of containment and surveillance where materials accountability alone proves to be less than fully adequate. What is even more important, however, is the imperative of clarifying the role of safeguards—where they fit in the broader scheme of things—and avoiding the perpetuation of misconceptions that lead to public overexpectation and inevitable loss of confidence.

Safeguards Improvements

Earlier discussion made clear that safeguards stand to be improved in a number of areas, ranging from developing and deploying more durable and sophisticated equipment, to improved methodologies for dealing with particular kinds of nuclear facilities, to enhanced procedures for processing and evaluating information and for assessing the character of anomalous events that might signal diversionary activity. These do not require detailed restatement here. Two points do deserve special emphasis however: the importance of national cooperation to safeguards effectiveness and the value of safeguards transparency.

For all of the technical detail surrounding safeguards, it is clear that they are fundamentally a political phenomenon. Their acceptance by states is a political statement of willingness to submit solemn undertakings to international scrutiny and verification. The technical elegance of sophisticated methodologies and procedures shades into insignificance when compared with the readiness of sovereign states to ensure that whatever is necessary to achieve safeguards effectiveness will be done. Cooperation, in the final analysis, is perhaps the single most important element in safeguards credibility.

The reality of safeguards implementation unfortunately often falls short of the mark on the matter of cooperation. States often incline toward narrow construction of safeguards agreements and subsidiary arrangements as, for example, when confronted with adapting safeguards procedures to accommodate new instrumentation and methodologies. In such circumstances, they often focus more on legalistic interpretation of their agreements than on the practical question of what is necessary and reasonable to ensure that the objective of independent verification can be effectively met. The problem is not one sided. For its part, the agency, like any large organization, often tends to retreat behind the protective shield of bureaucratic rules when confronted with difficult situations.

Cooperation is a two-way street requiring that states avoid unnecessary appeal to legalistic solutions while the agency avoids kneejerk invocation of bureaucratic rules; both must work toward practical solutions that upgrade the common interest of effective and credible safeguards. This admonition applies to such matters as avoiding doctrinaire positions regarding deployment of newly developed instrumentation, or facilitating the application of safeguards measures and approaches in a way that protects legitimate proprietary interests and avoids undue intrusion in the economic operation of nuclear facilities; it also applies to such measures as ensuring that steps are taken to improve state systems of accounting for and controlling nuclear materials in the interest of achieving timely reporting and comprehensiveness of records. Moreover, it extends to the willingness of states to serve as showcases to demonstrate new safeguards techniques and approaches and to test whether and to what extent they intrude on normal and efficient operations. States should also be willing to develop appropriate safeguards measures and instruments to deal with technologies that they may have been responsible for introducing into the nuclear marketplace. Canada's development of fuel bundle counters to deal with problems associated with safeguarding heavy water reactors is a case in point. In short, safeguards should be implemented with a spirit of cooperation and not merely in accordance with agreed legal requirements.

Safeguards systems are information systems, providing information regarding state behavior and serving as a basis for policy in states that factor that information into their national security planning. In democratic societies, where public opinion plays an important role in helping to shape public policy, the public also assimilates that information. One of the difficulties in reaching informed judgments about state behavior elsewhere is the extent to which information is regarded as private, privileged, and unavailable for public consideration. Problems related to confidence in the findings of the international safeguards system would seem to dictate the need for reassessment of how to deal with the information base on which assessments of compliance and the absence of diversion are made. This is especially appropriate in view of the increase in the past five years in public testimony by acknowledged experts and former IAEA inspectors regarding alleged weaknesses and deficiencies in the system.[25]

At least two kinds of information are involved: information at the state level, and information at the agency level, that is to say, information regarding the basic verification activities the agency uses to make its independent assessment of state compliance. The statute of the agency and the basic safeguards documents afford the state protection of privileged, proprietary, and commercial information, but do not extend this protection to the agency itself.

Bearing the statutory provisions in mind, consideration should be given to the feasibility of creating greater transparency of agency verification information. If the agency is embarrassed as a consequence of revealed deficiencies or weaknesses, this should be regarded and defended as part of the learning process and the evolution of safeguards systems. But consideration also should be given to offering countries, on a systematic basis, the opportunity to have the results of agency inspections publicized subject to the deletion of sensitive or proprietary material. There is virtually no proprietary information remaining in the realm of light water reactors for example, and that is probably true with respect to the vast majority of research reactors currently under safeguards. A "graded transparency" system might thus be introduced, even on a trial basis, in order to ascertain the extent to which such an arrangement would enhance confidence in verification activities and safeguards findings, and to identify possible problems posed for the inspected states. States with priority commitments to international safeguards might offer leadership by example, volunteering greater transparency with regard to the arrangements for and results flowing from safeguards activities. Here again is an opportunity for a constructive U.S. initiative.

Supplementary Institutions

The discussion of safeguards has emphasized the importance of not attributing purposes to them that they were neither intended nor equipped to fulfill. Safeguards can verify peaceful use of nuclear materials; they cannot protect them or physically prevent their diversion or seizure. Nor can they limit the location where sensitive materials will be produced or stored. But these are real policy issues and the source of concern for many of those who have questioned the efficacy of safeguards. They raise policy questions whose answers may include the adoption of additional requirements for international nuclear commerce and cooperation or the establishment of supplementary institutions. A number of such institutions have been identified in the past, none of which at the time was sufficiently attractive that it earned the support needed to be established. These include regional and multinational nuclear fuel cycle centers that might house enrichment or reprocessing plants or both, thereby presumably limiting the number of nationally owned and operated facilities producing sensitive nuclear materials; nuclear fuel banks that could provide assurance of adequate and timely supplies of natural or low enriched uranium, thereby reducing incentives for states to develop independent enrichment capacity; spent nuclear fuel storage facilities to alleviate pressure on national storage sites (a problem that, if unresolved, could adversely affect the efficient and economic operation of nuclear power plants as well as avoiding early or unnecessary reprocessing and storage of separated plutonium); and international plutonium storage.[26] Only the latter institution, which many regard as one of the most significant from a nonproliferation point of view, will be dealt with here.

Increased concern in the late 1970s over the proliferation risks associated with plutonium and reprocessing led to the establishment in 1978 of an IAEA expert group on international plutonium storage

(IPS). The notion of exploring such an arrangement also was endorsed by the International Nuclear Fuel Cycle Evaluation. Unlike several other suggested institutional approaches, IPS had a statutory antecedent in a previously unimplemented provision in Article XII.A.5 of the IAEA statute that contemplated that the agency safeguards system would include a right of the agency:

> to require that special fissionable materials recovered or produced as a by-product be used for peaceful purposes under continuing Agency safeguards for research or in reactors . . . and to require deposit with the Agency of any excess of any special fissionable materials recovered or produced as a by-product over what is needed for the above stated uses in order to prevent stock-piling of these materials, provided that thereafter at the request of the member or members concerned special fissionable materials so deposited with the Agency shall be returned promptly to the member or members concerned for use under the same provisions as stated above.

Those most interested in the potential nonproliferation benefits of such a system saw it as a way of avoiding national stockpiles of separated plutonium that could be diverted to use in weapons and of strengthening IAEA safeguards by adding a requirement that the peaceful uses to be made of separated plutonium be immediate ones, declared by the user, and giving the agency responsibility for verifying that the plutonium was in fact dedicated exclusively to such declared uses. Others, on the other hand, were concerned that such a system might be used to restrict their freedom of choice in the peaceful uses of plutonium and impose conditions (such as preclusion of any nuclear explosive use) to which they had not previously agreed.[27]

Alternative conceptual arrangements for implementing the principle of an IPS were reviewed over a period of five years, resulting in a report in November 1982 to the board of governors.[28] Differences had emerged during that period over the determination of what should constitute "excess" plutonium that would have to be deposited; how rigorous the requirements and decision-making arrangements for release of the plutonium from international custody should be; what rights would be adequate for the agency to effectively verify continued dedication of released material to the declared uses; and where plutonium stores should be located. On the other hand, there was general agreement that an IPS should be treated as part of the IAEA safeguards system and not as a new or separate system, and that established safeguards reporting and inspection procedures should be utilized to the extent possible. It also was generally recognized that it would be impracticable and unnecessarily costly for the agency to build its own plutonium storage facility and that IPS facilities therefore should be created by designating an existing facility as an international store under an appropriate agreement, covering custody and supervision, between the agency and the host state.

Differences in view outweighed points of concurrence; in the final analysis, the study group's report to the board of governors in 1982 could do no more than present three alternative schemes of varying degrees of rigor. One provided for a purely voluntary arrangement that would leave to the state the decision on what should be treated as "excess" plutonium as well as the right of return on request with no other conditions than to keep it under IAEA safeguards. The most rigorous alternative required that all separated plutonium in a state other than that required for immediate specified peaceful non-explosive uses must be registered and deposited in an approved international storage repository. In addition, it provided that release of such material would require a statement of intended use and for an IAEA assessment of that statement. A negative assessment could lead to refusal to release the material from international custody.

The board of governors took note of the report in February 1983, stating its intention to consider the matter further in the future. But it has yet to authorize any actual steps to implement an international plutonium storage system. Meanwhile, some rudimentary elements of a regional plutonium storage system were incorporated in a November 1984 Declaration of Common Policy on the Consequences of the Adoption of the London Guidelines by Ten Member States of the European Community,[29] and the Final Declaration of the 1985 NPT Review Conference reconfirmed international interest in some form of international plutonium storage arrangement in urging that the "IAEA establish an internationally agreed effective system of international plutonium storage in accordance with Article XII.A.5 of its Statute."[30]

Some countries, in particular the United States at one point, have discouraged establishment of an IPS out of concern that its existence might unnecessarily stimulate and legitimatize reprocessing and force the world into dealing with a plutonium economy without adequate protective institutional arrangements. Fortunately, demand for reprocessing either for waste management or as a fuel supplement has been almost nonexistent in recent years, as electricity demand and nuclear growth have slowed and economically competitive and abundant uranium resources have appeared in the marketplace. But any resurgence in nuclear demand could quickly alter this situation and confront the world once more with the problem of managing fuel discharges and ensuring adequate and economically competitive supplies of nuclear fuel.

Reprocessing, of course, is not the only or necessarily the preferable waste management strategy. The reality, however, is that some states seem determined to keep it on their agenda. Separated plutonium, which is today confined almost exclusively to nuclear weapon states and several advanced industrial states (India is an exception) could eventually become more widespread. While an IPS does not solve the plutonium problem or remove concern about the technical limits of material accountancy safeguards in large reprocessing plants, it could, if appropriately designed, provide reinforcement to safeguards and to the nonproliferation regime.

Even while making every effort to avoid the presence of separated plutonium beyond what is absolutely essential to peaceful nuclear activity, and while searching for alternative approaches to plutonium for extending the energy value of the current nuclear fuel cycle, it is important to seek to establish collateral institutions to increase social controls over potentially dangerous activities and materials. It is already reasonably clear that conventional safeguard measures will always leave a residue of uncertainty in some areas of the nuclear fuel cycle. It is also clear that nations will never be fully certain about the future intentions of their neighbors and that no safeguards systems alone could (or should be expected to) provide that assurance. This suggests the need for eternal vigilance in respect of nuclear activity. But it also calls attention to institutional strategies that can reinforce verification safeguards, further reduce the risk of diversion, complicate access to dangerous nuclear materials, and make more difficult the political choice for would-be denunciators of nonproliferation undertakings. Institutional approaches deserve particularly close attention if, in the course of shoring up nonproliferation, they also assist in improving predictability of supply and in strengthening international nuclear cooperation.[31]

Depoliticization

Proposition. The intrusion of some politics into the life of the IAEA is unavoidable. What must be guarded against is allowing the agency to become dominated by extraneous political issues that can contribute to polarization, undermine its ability to effectively carry out its responsibilities, and generally weaken confidence in the organization. Achieving this objective is an important challenge to agency leadership. The best chance of meeting it depends on building a strong consensus against the introduction into agency deliberations of extraneous political issues. This in turn depends on the existence of a coalition of states representing all of the principal membership groups in the agency, each of whom has a strong stake in agency stability and effectiveness. This requires acknowledging the diversity of interests among the agency's members, ensuring that those interests are adequately reflected in the agency's programmatic agenda, and that states do not seek to treat the agency simply as an extension of their own national institutions and interests. Complete depoliticization is not a feasible goal, but it can be somewhat approximated if a policy of genuine universality is practiced by all of its membership in respect to all phases of agency activity. Countries like the United States, which have a very substantial interest in the international safeguards administered by the IAEA, have a special responsibility to avoid letting other national interests intrude on their commitment to the agency.

Argument. On the matter of depoliticization, two issues are clear. First, continued intrusion of extraneous political issues into agency activities threatens to undermine confidence of the United States and other major member states in the IAEA's ability to effectively carry out its responsibilities. Second, it is unrealistic to think that politics can be entirely excluded from the agency, and, indeed, it would be historically inaccurate to claim that the agency was ever totally nonpolitical.[32]

At a minimum, *intrinsic* politics are inherent in any organization, and this is likely to occur even more frequently in an organization such as the IAEA that touches directly the sensitive nerve of national sovereignty. As for *extraneous* politics, its total exclusion, while desirable, is almost certainly not achievable. In an organization populated by heterogeneous groups of states reflecting a diversity of interests and opinions and differing priorities, it is inevitable that political issues will creep in. Not all such political issues necessarily contribute to or reinforce polarization unless those who do not share the interests or concerns underlying those issues choose, by the nature of this reaction, to so interpret them.

States, then, must learn to live with at least a degree of extraneous politicization and to focus attention on mitigating its impact on the main purposes and the effectiveness of the organization. To the extent possible, the introduction of such issues should be curtailed, but where this cannot be achieved, the main emphasis must be on isolating the issue and avoiding its linkage to other issues and activities that are legitimately part of the agency's agenda and of its statutory responsibility. It is in the nature of things that that burden falls disproportionately on states whose interests are most closely reflected in the prevailing principles and norms of the organization and who, therefore, have the greatest stake in avoiding disruption. This is a challenge of artful diplomacy and of self-discipline in dealing with occasionally troublesome and even disabling behavior and attitudes.

Defining *extraneous* politics is less simple than it might appear. The United States considered the Israeli attack on Iraq's safeguarded research reactor to be an extraneous and therefore inappropriate issue for discussion in the committees and forums of the IAEA. Many other states—including a number of Western states—did not share that view. This simple difference underscores that to define excludable issues, or, if agreement cannot be reached, to handle them so as to minimize the risk of their becoming unmanageable and destructive, is no easy task. Avoiding the trap of playing the same politicization game one seeks to foreclose is also not easy. There is a strong case to be made that the United States, while quite properly defending the agency's statute by resisting efforts to restrict the rights of particular members to participate in the IAEA, in the final analysis let its support for Israel get in the way of defining a middle ground that might have defused the Israeli issue in the agency, and that the U.S. response actually increased the level of politicization and contributed to its perpetuation.

A number of U.S. allies were surprised and dismayed that the United States appeared willing to risk the collapse of the safeguards system over the issue of the treatment of Israel, and for the first time in the agency's history were led to contemplate the depth of U.S. commitment to an agency-anchored nonproliferation regime. This suggests the need for increased self-discipline on the part of the United States: while it should make clear that notwithstanding its commitment to the IAEA it cannot allow that commitment to be held hostage to the protection of other crucial national interests, it must do so without threatening to withdraw whenever contentious political issues arise. Extraneous politicization is an unquestionably difficult problem with no easy or obvious solutions, but absolutist approaches often create more problems than they solve. Europeans may adjust more easily to the realities of politicization than do Americans, and also may be more disposed toward liberally defining what is reasonable and necessary in an effort to avoid frustrating have-not nations who are impatient to advance their objectives, and who sometimes seek to take advantage of what they see as advanced country interests to press extraneous claims.[33]

Successful depoliticization depends on many factors. One, as suggested by the preceding discussion, is to build a consensus not only on what constitutes extraneous issues but also on the unacceptability of their introduction in the deliberations and activities of the agency. This cannot be achieved by a small group of advanced states alone. It entails reaching out to Third World nations to bring the moderates among them into the consensus-building process from a very early stage. This, unfortunately, has not been the typical practice, and a perceived insufficiency in early and meaningful consultation has contributed to continued Third World pressures to increase their representation on the board of governors. Moreover, the Third World suspects that the consensus-decision process, to which they have not been party, "pre-cooks" many decisions and that their only recourse is to gain numerical control of the board and to change decision making from consensus to formal voting. That prospect alone should provide an incentive to alter the consensus-building structure to include the moderate developing countries.

Presuming that the Israeli issue has indeed been brought under control and will not continue to pervade agency activity in the future as it had during a period of five years, the door is open for a return to normalcy.[34] But normalcy must be defined against the political realities of the 1980s, *not* those of the 1960s. Over time, the United States and other advanced countries have become accustomed to a certain behavioral and policy-making pattern and atmosphere in the IAEA. A tendency also has developed to regard the agency as an extension of U.S. or U.S.-Soviet policy with major emphasis on safeguards. This view clearly is not shared by all IAEA members, some of whom see safeguards primarily in terms of a price to be paid for technical assistance; that is, as a cost, not a benefit. The United States firmly adheres to the principle that safeguards are in the interest of all and that all, therefore, should share to some degree in contributing to their cost. It also has strongly asserted the principle of universality, especially when dealing with the Israeli question.

As reasonable and well-intentioned as this position may be, it bears repeating that the viability of the agency depends on accommodation of an array of interests, and a recognition of another kind of universality—one that accepts and respects the differences of priority and perspective on the part of developed and developing countries, of member states that are party to the NPT and those that are not, of consumers as well as suppliers, and of East and West. There is a need to enlist the support of all of these groups in reinforcing and maintaining a consensus that is sound and resilient enough to withstand, if not prevent, aberrant events and challenges to the IAEA's integrity. At the same time, it is necessary to assure a broad commitment to, and necessary support for, a safeguards system that has the confidence of the international community. The costs of that system, which benefit all, should be shared by all. Strong U.S. support for the agency, coupled with active efforts to work closely and on a timely basis with moderate developing states on programmatic, safeguards, resource allocation, and general policy issues, could well keep inevitable differences of interest under control and minimize the risk of disruptive and unrelated political issues undermining agency effectiveness and credibility.

The Balance Between Safeguards and Technical Assistance

Proposition. The health and robustness of the agency depend in large measure on the depth of commitment of a cross-section of its total membership. The objective should be the existence of a clientele for whom weakening of the organizations would be costly or even unacceptable. Balance is important to the generation and maintenance of strong membership stakes in an effective organization that is not diverted from its central responsibilities. The assurance of such balance, therefore, should be a priority concern of agency leadership. An effort should be made to provide a degree of predictability in the level and availability of technical cooperation resources that will engender a confidence and a stake that will serve to reinforce the agency's foundations. Commitment to basing some portion of technical cooperation funds on assessments could be an important vehicle for generating the confidence and the stake. Even more important is assurance that the levels of support in a balanced program are adequate to permit the agency to achieve the objectives expected of it. This requires moving beyond the current zero-growth posture.

Argument. In his provocative analysis of international safeguards, David Fischer remarks that the real Achilles' heel of the NPT is the large number of primarily Third World states that believe that they would stand to lose very little by leaving the treaty.[35] That observation has particular significance for the IAEA, which counts among its members a number of states that have little if any nuclear activity or infrastructure and that, as we have noted, are rather susceptible to being mobilized by activist states with agendas that are not always in the broader interests of the agency. It is a matter of some concern, then, to avoid the risk that a group of member states might come to view the undermining or weakening of the agency with indifference.

Members must have important stakes in the robustness of the IAEA. We have commented more than once on the insistence of the United States and some others that safeguards are in everybody's interest and must be protected and reinforced in any reasonable way.

This belief underlies the insistence that no matter how modest it may be, every member state must contribute to the cost of maintaining international safeguards. The importance of safeguards may be incontestable—indeed, that is a fundamental assumption of this book—but not all states necessarily appreciate this or support this view with the same degree of enthusiasm and conviction as those responsible for the establishment and perpetuation of safeguards, the strong endorsement of the Third NPT Review Conference notwithstanding. There is every reason to continue to intone the virtues and importance of safeguards, but it also is necessary to cultivate each member state's commitment to the vitality and effectiveness of the IAEA. Moreover, it is vital that all groups of states, acting as cohesive political units, be persuaded of the IAEA's importance to their collective interests.

Technical assistance and cooperation are highly valued; indeed, they are the prime concern of a large segment of the agency's membership. The extent of developing country support for the interest of advanced states in safeguards is in no small measure dependent on the sensitivity and attention of advanced states to developing countries' interests. It is not just a question of numbers of dollars. Of course, developing countries are keenly sensitive to how the balance sheet of safeguards and technical assistance allocations appears and whether increases in safeguards allocations are significantly disproportionate to the amount of resources made available for technical assistance activities. This concern is reflected in the recommendation of the Third NPT Review Conference that the IAEA be provided with the "necessary financial and human resources to ensure that the Agency is able to continue to meet effectively its responsibilities."[36] But developing countries place equal importance on the degree of certainty and predictability of resources for technical assistance and on the depth of commitment of advanced industrial states to ensure that adequate and timely assistance is available on a continuing basis to meet legitimate Third World requirements.

The introduction and continued practice of adopting indicative planning figures has gone far toward relieving some of the concerns of the relatively recent past. The indicative planning approach will continue through 1988 and very possibly will be renewed. However, for many states this does not fully resolve the problem of placing technical assistance on a more certain and predictable basis for the long run.[37] Technical assistance traditionally has been funded through voluntary contributions, whereas safeguards funding is based on assessments. Since the issue has a substantial politically symbolic content, one approach would be to adopt a floor of assessed technical assistance (on the order of 15 to 20 percent) with agreement that, in addition to the amount assessed, the principle of indicative planning figures would continue to be used and that growth levels would remain reasonably commensurate with identified need. Although this would modify the voluntary basis on which technical assistance currently is funded, it would do so in only a limited way. What is important is that in doing so it would create a political visibility for technical assistance that could remove a source of irritation, deepen a country's stake in the agency, and enhance overall support.

There is also a need to avoid behavior that generates uncertainty and doubt. Withholding resources (as the U.S. Congress on several occasions has done in the aftermath of the general conference rejection of Israeli credentials in 1982) sends the wrong message: it implies both a lack of concern with the technical assistance so highly valued by developing countries and a lack of commitment to the purposes of the agency when the agency plays a role that is central to our national security interests and to the global stability we seek. Whatever short-term effect such action might have for United States or Israeli objectives, it is highly doubtful that it could have any long-run benefit. Indeed, since other countries gauge their support and enthusiasm for the agency at least partly in terms of U.S. behavior, our actions tend to have spillover effects that can magnify costs to the IAEA. That is counterproductive.

This point is also relevant at the broader level of assurance of nuclear supply. We may recall from chapter 6 that in the late 1970s a major source of discontent among developing countries was a tightening of nuclear export policies by leading nuclear supplier states. One reason for this discontent was the adoption of the Nuclear Supplier Guidelines after consultations in which the developing countries had not been invited to participate; another was the subsequent even further unilateral tightening of supply conditions by several major suppliers (including the United States, Canada, and Australia). Efforts to open this subject to broad international discussion included the International Nuclear Fuel Cycle Evaluation of 1978–1980, and the establishment in 1980 of the IAEA Committee on Assurances of Supply (CAS) to consider and advise the board of governors on predictable long-term supply assurances consistent with effective assurance of nonproliferation.

As noted in chapter 3, CAS has thus far achieved a number of positive results.[38] An emergency back-up mechanism has been worked out to cover interruptions or delays in the provision of contracted supplies that are not caused by a breach of agreement by the recipient and that are likely to lead to a disruption or delay in the construction or operation of a nuclear facility in a state that is in compliance with its nonproliferation and safeguards commitments. In addition, modest, but not insignificant, amendment procedures for the revision of intergovernmental nuclear cooperation agreements have been identified and defined; and measures for alleviating technical and administrative problems in international shipments of nuclear materials and equipment have been completed.

Implementation of these measures, however, is largely contingent on completion of a set of principles of international cooperation in the field of nuclear energy. Progress toward this end has been slow but has narrowed to the question of achieving agreement on what form nonproliferation commitments should take, and on the scope of agency safeguards related to those commitments. The different views represented in the CAS reflect the different perspectives on nonproliferation noted earlier: for some, unilateral assurances of nonproliferation should suffice; for others, such assurances only can be effective if inscribed in multilateral treaties such as the NPT or Treaty of Tlatelolco—in short, differences over how nonproliferation should be defined; how it should be related to assurance of supply; and just what the nature of the nonproliferation commitment should be.

The broad differences between NPT and non-NPT states on such questions suggest that the challenge is one of reconciling the irreconcilable. Nevertheless, the CAS is the preferred place to make this effort, to seek some political accommodation of views that will permit implementation of other agreed mechanisms related to security of supply and that will provide a point of departure for further discussion of the issues involved. It would be unfortunate if the United Nations Conference for the Promotion of International Cooperation in the Peaceful Uses of Nuclear Energy (PUNE) were to convene in 1987 in the absence of a set of baseline principles derived in CAS, because it is implausible to think that in such a situation PUNE would not focus its energy and attention on the very same questions but this time in a highly publicized forum that would lend itself to the very kind of political recrimination and hostility that the world nuclear community should avoid. For PUNE to play the desired constructive role, it is essential that it be launched in reasonably settled circumstances, drawing full benefit from the IAEA's extensive efforts during the past half decade to stabilize the conditions of international nuclear cooperation. This is another challenge for leadership, but one that in the final analysis can be successfully met if advanced and developing, NPT and non-NPT states alike, share the burden.[39]

A FINAL JUDGMENT

All of these suggestions and improvements aside, we cannot leave this subject without recalling a point made at the outset, namely that proliferation is, in the final analysis, a political question. Whether states are tempted to seek to divert nuclear material or to breach or withdraw from solemn obligations not to seek to acquire nuclear weapons depends upon fundamental political judgments they make about their security, their international status, related political matters including domestic political considerations, and how, from their perspective, all of these elements relate to nuclear weapons.

What the superpowers do or fail to do in the arena of arms control is part of that decision-making fabric, as is their political behavior in the world and the effect that it has on the security and political interests of different states. This is not to say that a decision by the superpowers to bring a halt to the vertical arms race would necessarily foreclose a non-nuclear weapon state from making a fateful, political decision to acquire nuclear arms. But the superpowers' continued development of nuclear arsenals in terms of numbers or level of sophistication surely undermines their moral basis for arguing against nuclear weapons for others, and can in this manner weaken the fabric of the nonproliferation regime.

The regime is alive and reasonably well today, enjoying the afterglow of what by all accounts was a successful Third NPT Review Conference in 1985. But alive and well does not mean that the regime is not fragile or that it cannot be fractured. It is not simply a question of hardware, but also of doctrine and perception. Political decisions reflect objective realities filtered through the lenses of perception. Efforts to shape those perceptions in the direction of delegitimizing nuclear weaponry *everywhere* is a condition precedent to successfully delegitimizing them *anywhere*. That reality imposes substantial responsibilities on the nuclear leviathans—a responsibility that, if ignored, could contribute significantly to international instability.

This being said, we can return to the main theme of our study. Nuclear energy is with us indefinitely. There are today 397 nuclear power reactors in operation in 26 states, accounting for as much electricity as the entire world produced from all sources in 1955. Another 85 reactors are under construction in these and 7 other states. Another 335 research reactors of all kinds dot the global landscape with more under construction or in various stages of planning. The cumulative plutonium content of spent fuel discharged in nonmilitary activities to date is estimated to be about 210 tons, about

7 tons of which are in separated form. Pandora's box is open. The technology cannot be disinvented. The challenge now is to keep it under political, social, and technical controls so that it serves the interest of mankind and does not destroy it. That is a challenge that requires the collective effort of mankind working together in common ways and institutions.

One such institution already has been fashioned, the International Atomic Energy Agency. Among international organizations, it has an enviable record of achievement. It has been vested with responsibilities and authority that, however constrained they may be, reach beyond that with which any other contemporary global international institution has been endowed. To appreciate the confidence it has earned in the international community, one need look no further than the scope of safeguards responsibilities with which it has been charged, the strong endorsement it received at the Third NPT Review Conference, the investment of political, economic, and human capital the United States has made in the organization, and Soviet General Secretary Gorbachev's May 1986 public appeal to "enhance the role and possibilities of that unique international organization." The circumstances giving rise to that appeal have created an opportunity few anticipated, a window of opportunity for consolidating the gains the agency has made and for providing it with the resources to serve the international community in whatever way the latter in its collective wisdom sees fit. It is a rock on which to build, a foundation for civilization's future. To forge ahead successfully requires unstinting support from the nations of the world, a shedding of the conceptual blinders of sovereignty, indeed an understanding that an effective and credible international atomic energy organization enhances their own national sovereignty by providing an added dimension of certainty and security to the environment. If the risk of proliferation has not already made this abundantly clear, the tragedy surrounding Chernobyl surely must.

NOTES

1. For the first two periods under review, see Lawrence Scheinman, "IAEA: Atomic Condominium?" in R. W. Cox and H. K. Jacobson, eds., *The Anatomy of Influence: Decision Making in International Organization* (New Haven and London, Yale University Press, 1973) and Alan McKnight, *Atomic Safeguards: A Study in International Verification* (New York, UNITAR, 1971).

2. Efforts to restructure the international economy and the agenda of the South are discussed in Stephen D. Krasner, *Structural Conflict: The Third World Against Global Liberalism* (Berkeley, University of California Press, 1985).

3. The origins and early development of UNCTAD are analyzed in depth in Joseph S. Nye, "UNCTAD: Poor Nations' Pressure Group" in Cox and Jacobson, eds., *Anatomy of Influence* (see note 1 above).

4. Majoritarianism refers to the practice of bloc voting in pursuit of special interests and with a view to pressing an agenda regardless of the possible long-term costs to the institution in which voting occurs.

5. The guidelines are contained in IAEA document INFCIRC/254 (IAEA, Vienna, 1978). See especially paragraphs 2 and 4 of that document. The guidelines were agreed upon by participants in the Nuclear Suppliers Group: Belgium, Canada, Czechoslovakia, France, the German Democratic Republic, the Federal Republic of Germany, Italy, Japan, Netherlands, Poland, Sweden, Switzerland, United Kingdom, United States, and the Union of Soviet Socialist Republics.

6. This position is well outlined by one of the Third World participants, Munir Ahmad Khan, in "Nuclear Energy and International Cooperation: A Third World Perception of the Erosion of Confidence," Working Paper of the International Consultative Group on Nuclear Energy (New York and London, The Rockefeller Foundation and the Royal Institute of International Affairs, September 1979).

7. Significantly, the general conference met for its twentieth session that year in Rio de Janeiro, Brazil.

8. See IAEA, GC(XXIX)/RES/443 (September 27, 1985).

9. The issue arose again at the thirtieth general conference held in October 1986 when a group of Islamic states introduced a resolution focusing on Israel as a Middle East nuclear threat rather than on the Israeli attack on the Iraqi reactor in 1981. The operative paragraphs called for establishment of a Middle East nuclear-weapon-free zone and for Israel placing all of its nuclear facilities under IAEA safeguards. It also called on all states to discontinue cooperation with Israel in the nuclear field until compliance with the provisions of the draft resolution had been achieved. A series of procedural measures resulted in an adjournment of the debate on this resolution; no action was taken. Significantly, the Soviet bloc for a second year in a row has abstained when punitive measures have been called for against Israel suggesting significant Soviet interest in stabilizing the political situation in the agency. It is nevertheless still unclear whether 1987 will see another attempt by the Arab states to isolate Israel in the agency. The debate at the 1986 general conference can be followed in IAEA, GC(XXX)/OR.20 (October 2, 1986).

10. For an overview of this frame of mind, see Bertrand Goldschmidt and Myron R. Kratzer, "Peaceful Nuclear Relations: A Study of the Creation and Erosion of Confidence," in Ian Smart, ed., *World Nuclear Energy: Toward a*

Bargain of Confidence (Baltimore, Johns Hopkins University Press, 1982) pp. 19–48.

11. This case is argued most effectively in Victor Gilinsky, "Plutonium, Proliferation and Policy," *Technology Review* vol. 79, February 1977, pp. 58–65. See also his "Plutonium, Proliferation and the Price of Reprocessing," *Foreign Affairs* vol. 57 (Winter 1978–1979) pp. 374–386. See also Albert Wohlstetter and coauthors, *Swords from Ploughshares: The Military Potential of Civilian Nuclear Energy* (Chicago, University of Chicago Press, 1979).

12. This line of argument is fairly persistent in the general conference discussions of the past decade.

13. It is equally difficult to persuade governments that by creating greater international authority they actually may increase their own ability to more effectively control activities under their jurisdiction by virtue of having the benefit of authoritative international rules and norms to constrain the behavior of others.

14. The Convention on Early Notification of a Nuclear Accident describes the IAEA as a focal point for information and notification. The Convention on Assistance in the Case of a Nuclear Accident or Radiological Emergency sees the IAEA's charge as being to collect and disseminate information concerning experts, equipment, or materials that could be made available in case of an emergency and to assist in preparing emergency plans, developing training programs for dealing with nuclear accidents, maintaining liaison with other relevant international organizations for the purpose of obtaining and exchanging relevant information, and similar activities. In addition, the board of governors in September 1986 approved an augmented safety program for the agency for the next several years.

15. The text of the statement made on May 14, 1986, by General Secretary M. S. Gorbachev is published in IAEA, GOV/INF/497, May 20, 1986.

16. See *The New York Times*, June 4, 1986, for a report on the message from General Secretary Gorbachev.

17. See Review Conference of the Parties to the Treaty on the Non-Proliferation of Nuclear Weapons, *Final Document*, Part I, Final Declaration relative to Article III and preambular paragraphs 4 and 5, paragraphs 12, 5, and 21 respectively. NPT/CONF.III/64/I (Geneva, IAEA, 1985) Annex I, pp. 2–6.

18. In taking this position, we by no means ignore the likelihood that the Soviet Union sees in this course of action an opportunity to dilute their own problem by emphasizing the generic nature of the nuclear safety issue and that Chernobyl is but one of a series of accidents that have occurred in the past in the United States and United Kingdom. What we do say is that it is in the U.S. interest to capitalize on the Soviet's willingness to strengthen the IAEA since this is fully consistent with U.S. national interests. Some U.S. officials have seen Chernobyl and its aftermath as an opportunity to embarrass and needle the Soviets rather than as an opportunity to advance our own national interests.

19. "The International Atomic Energy Agency: An Appraisal with Recommendations for United States Policy," A Report to the Secretary of State of the Panel to Review the Activities of the International Atomic Energy Agency (1972). This panel was chaired by Allen V. Astin, Director Emeritus of the National Bureau of Standards and included five highly qualified private citizens. The report is reprinted in *Nuclear Safeguards: A Reader*, A report prepared by the Congressional Research Service, Library of Congress, for the Subcommittee on Energy Research and Production, transmitted to the Committee on Science and Technology, U.S. House of Representatives 98 Cong. 1 sess. (Washington, D.C., U.S. Congress, December 1983) p. 108. (Hereafter referred to as CRS, *Nuclear Safeguards Reader*.)

20. IAEA, INFCIRC/153, paragraph 28.

21. The agency has developed comprehensive criteria for self-evaluation of how well it is progressing in the application of safeguards. But the criteria are extensive and complex and not easily translated into simple criteria that could be used by political decision-makers to assess in clear and concise terms the effectiveness of IAEA safeguards. That task still remains.

22. The most comprehensive argument in the public domain is André Petit, "De la necessite d'approfondir le consensus international quant aux implications pratiques des concepts de base des garanties internationales," a paper presented at the Sixth ESARDA Symposium on Safeguards and Nuclear Material Management, Venice, Italy, May 14–18, 1984. The paper is printed in the Proceedings of the ESARDA Symposium and reprinted in Aspen Institute for Humanistic Studies, *Proliferation, Politics and the IAEA: The Issue of Nuclear Safeguards*, Part II (Berlin, June 1985).

23. See comments of André Petit at Aspen Institute Berlin Conference cited in note 22 above.

24. See *Final Document*, Non-Proliferation Treaty Review Conference, Final Declaration relative to Article III and preambular paragraphs 4 and 5, and paragraph 5 (see note 17 above).

25. See, for example, the testimony of two former IAEA inspectors, Roger Richter and Emmanual Morgan, at hearings in Congress in the aftermath of the Israeli attack on Iraq's nuclear reactor in June 1981. U.S. Congress, Senate Committee on Foreign Relations Hearings on the Israeli air strike 97 Cong. 1 sess. (June 18, 19, and 25, 1981). Richter's statement is reproduced in CRS, *Nuclear Safeguards Reader* (see note 19 above) p. 719; and hearings on IAEA programs of safeguards before the Senate Committee on Foreign Relations, 97 Cong. 1 sess. (December 2, 1981). Morgan's statement is reproduced in CRS, *Nuclear Safeguards Reader*, p. 748.

26. For in-depth analyses of the multinational approach see Myron B. Kratzer, *Multinational Institutions and Non-proliferation: A New Look*, Occasional Paper No. 20 (Muscatine, Iowa, The Stanley Foundation, 1979); and Lawrence Scheinman, "Multinational Alternatives and Nuclear Nonproliferation," in George Quester, ed., *Nuclear Proliferation: Breaking the Chain* (Madison, University of Wisconsin Press, 1981) pp. 77–102.

27. It is important to bear in mind that (1) this gives the agency an authority, not an automatic right, and (2) it was drafted with the idea that the IAEA would be *the* supplier in many cases, a notion that turned out not to be correct. The discretionary authority applies only to agency projects; in other cases, voluntary participation would be needed to invoke the statutory authority. This is one reason why an IPS arrangement has not yet been made.

28. IAEA, Expert Group on International Plutonium Storage—Report to the Director General. IAEA-IPS/EG/140 (Rev.2) November 1, 1982. For an analysis of the discussions on IPS, see Charles N. Van Doren, "Toward an Effective International Plutonium Storage System," Report No. 81–255 S a report prepared under a research contract for the Congressional Research Service (Washington, D.C., CRS, November 1, 1981).

29. The declaration is reprinted in Appendix E of Council on Foreign Relations, *Blocking the Spread of Nuclear Weapons: American and European Perspectives* (New York, Council on Foreign Relations, 1986).

30. *Final Document,* Non-Proliferation Treaty Review Conference Final Declaration relative to Article III and preambular paragraphs 4 and 5, and paragraph 14 (see note 17 above).

31. Pursuing creation of an IPS to deal with civil plutonium produced anywhere does not remove the rationale, nor should it diminish efforts, to establish spent fuel storage arrangements that could help avoid the creation of "excess" separated plutonium that would result from reprocessing in advance, or in excess of, actual requirements for research or commercial purposes. Such a measure also would come to grips directly with the reality that for the foreseeable future the amount of spent fuel requiring storage will greatly exceed the amount that could be reprocessed. An expert group created under IAEA auspices in 1979 to study spent fuel management recognized that creation of an international storage center as an alternative or supplement to national storage offered potential practical advantages, but virtually no support for making use of such a mechanism mandatory, and this was reflected in the group's report of its findings to the board of governors in 1982. The NPT Review Conference considered spent fuel management to be primarily a national responsibility, but nevertheless endorsed the notion of international cooperation on the matter.

32. Within limits, it could be argued that a degree of politicization might even be healthy insofar as it keeps participants alert to the political realities of the environment in which they must function.

33. For a discussion along these lines, see Aspen Institute for Humanistic Studies, *Proliferation, Politics and the IAEA* (see note 22 above).

34. See note 9 above regarding this probability.

35. See David Fischer and Paul Szasz, *Safeguarding the Atom: A Critical Appraisal* (Stockholm, Stockholm International Peace Research Institute, Taylor and Francis, 1985).

36. See note 17 above.

37. At the thirtieth general conference held in October 1986, Belgium stated that it no longer wanted to participate in the indicative planning figure scheme as it could agree to technical assistance only as long as it was voluntary in character, and it would therefore decide on its technical cooperation contribution on a year-by-year basis. It would require only a few key defections of this nature to undermine the IPF approach. It may, therefore, become necessary in time to directly confront the assessment question.

38. For an overview of the CAS through mid-1983, see Charles N. Van Doren, "Nuclear Supply and Non-Proliferation: The IAEA Committee on Assurances of Supply," Report No. 83–202 S a report prepared under a research contract for the Congressional Research Service (Washington, D.C., CRS, October 1983).

39. UNCPIC PUNE convened as this book was going to press. The worst fears of a degenerative political debate did not materialize. But a conference that was nearly a decade in the making terminated with no substantive result. Valuable recommendations on ways and means of promoting international nuclear cooperation could not be adopted because no agreement could be reached on general principles of international cooperation, and many states in the G-77 felt that agreement on ways and means was contingent on agreement on principles. In the final analysis, UNCPIC PUNE reaffirmed continuing differences over the meaning of sovereignty, the meaning of non-proliferation, the meaning of non-discrimination, and the meaning of internationally binding commitments. While its adjournment closed one chapter of history, it also opened another in which the international nuclear community will have to contend once again with fundamental issues on whose positive resolution the future character of international nuclear cooperation and the non-proliferation regime will substantially depend.

APPENDIXES

APPENDIX A. THE MEMBERS OF THE AGENCY

On 1 August 1986 the 113 members of the agency were as follows:

Afghanistan
Albania
Algeria
Argentina
Australia
Austria

Bangladesh
Belgium
Bolivia
Brazil
Bulgaria
Burma
Byelorussian Soviet Socialist Republic

Cameroon
Canada
Chile
China
Colombia
Costa Rica
Côte d'Ivoire
Cuba
Cyprus
Czechoslovakia

Democratic Kampuchea
Democratic People's Republic of Korea
Denmark
Dominican Republic

Ecuador
Egypt
El Salvador
Ethiopia

Finland
France

Gabon
German Democratic Republic
Germany, Federal Republic of
Ghana
Greece
Guatemala

Haiti
Holy See
Hungary

Iceland
India
Indonesia
Iran, Islamic Republic of
Iraq
Ireland
Israel
Italy

Jamaica
Japan
Jordan

Kenya
Korea, Republic of
Kuwait

Lebanon
Liberia
Libyan Arab Jamahiriya
Liechtenstein
Luxembourg

Madagascar
Malaysia
Mali
Mauritius
Mexico
Monaco
Mongolia
Morocco

Namibia
Netherlands
New Zealand
Nicaragua
Niger
Nigeria
Norway

Pakistan
Panama
Paraguay
Peru
Philippines
Poland
Portugal

Qatar

Romania

Saudi Arabia
Senegal
Sierra Leone
Singapore
South Africa
Spain
Sri Lanka
Sudan
Sweden
Switzerland
Syrian Arab Republic

Thailand
Tunisia
Turkey

Uganda
Ukrainian Soviet Socialist Republic
Union of Soviet Socialist Republics
United Arab Emirates
United Kingdom of Great Britain and Northern Ireland
United Republic of Tanzania
United States of America
Uruguay

Venezuela
Viet Nam

Yugoslavia

Zaire
Zambia
Zimbabwe

Source: International Atomic Energy Agency, INFCIRC/2/Rev. 36, September 1986.

APPENDIX B. PARTIES[1] TO THE NUCLEAR NON-PROLIFERATION TREATY, INCLUDING 3 NWS, 9 DECEMBER 1986

Europe

1 Austria
2 Belgium
3 Bulgaria
4 Cyprus
5 CSSR
6 Denmark
7 Finland
8 German Dem. Rep.
9 Germany, Fed. Rep.
10 Greece
11 Holy See(acc)
12 Hungary
13 Iceland
14 Ireland
15 Italy
16 Liechtenstein(acc)
17 Luxembourg
18 Malta*
19 Netherlands
20 Norway
21 Poland
22 Portugal(acc)
23 Romania
24 San Marino*
25 Sweden
26 Switzerland
27 Turkey
28 USSR
29 UK
30 Yugoslavia

Asia and the Pacific

1 Afghanistan
2 Australia
3 Bangladesh(acc)
4 Brunei Darussalam*(acc)
5 Bhutan*(acc)
6 Taiwan, China*
7 Dem. Kampuchea(acc)
8 Fiji*(acc)
9 Indonesia
10 Japan
11 Korea, Rep. of
12 Lao People's Dem. Rep*
13 Malaysia
14 Maldives*
15 Mongolia
16 Nauru*
17 Nepal*
18 New Zealand
19 Papua New Guinea*(acc)
20 Philippines
21 Singapore
22 Solomon Islands*(succ)
23 Sri Lanka
24 Thailand(acc)
25 Tonga*(acc)
26 Tuvalu*(succ)
27 Viet Nam, Soc. Rep. of(acc)
28 Western Samoa*(acc)
29 Seychelles*(acc)
30 Kiribati*(succ)
31 Korea, Dem. People's Rep.(acc)

Africa and Middle East

1 Benin*
2 Botswana*
3 Burkina Faso*
4 Burundi*(acc)
5 Cape Verde*(acc)
6 Central African Rep.*(acc)
7 Chad*
8 Congo, People's Rep.*(acc)
9 Dem. Yemen*
10 Egypt
11 Equatorial Guinea*(acc)
12 Ethiopia
13 Gabon(acc)
14 Gambia*
15 Ghana
16 Guinea Bissau*(acc)
17 Iran
18 Iraq
19 Côte d'Ivoire
20 Jordan
21 Kenya
22 *Kuwait*[2]
23 Lebanon
24 Lesotho*
25 Liberia
26 Libyan Arab Jamahiriya
27 Madagascar
28 Mali, Rep. of
29 Mauritius
30 Morocco
31 Nigeria
32 Rwanda*(acc)
33 Senegal
34 Sierra Leone(acc)
35 Somalia*
36 Sudan
37 Swaziland*
38 Syrian Arab Rep.
39 Togo*
40 Tunisia
41 Uganda(acc)
42 United Rep. of Cameroon
43 Yemen Arab Rep.*
44 Zaire
45 Malawi*(succ)

The Americas

1 Antigua & Barbuda*(succ)
2 Bahamas*
3 Barbados*
4 Belize*(succ)
5 Bolivia
6 Canada
7 Colombia
8 Costa Rica
9 Dominica*(succ)
10 Dominican Rep.
11 Ecuador
12 El Salvador
13 Grenada*(acc)
14 Guatemala
15 Haiti
16 Honduras*
17 Jamaica
18 Mexico
19 Nicaragua
20 Panama
21 Paraguay
22 Peru
23 St. Lucia*(acc)
24 Suriname*(succ)
25 Trinidad and Tobago*
26 USA
27 Uruguay
28 Venezuela
29 St. Vincent & Grenadines*(succ)

Source: International Atomic Energy Agency, Vienna, Austria, December 1986.
*Non-member of the IAEA.

1. A state becomes a party to the NPT on deposit of an instrument of ratification, accession, or succession.

2. States italicized have signed but not yet ratified the NPT. States not italicized have ratified/acceded/succeeded.

APPENDIX C. CHRONOLOGY OF DEPOSITS OF RATIFICATIONS, ACCESSIONS, AND SUCCESSIONS TO THE NUCLEAR NON-PROLIFERATION TREATY

Non-nuclear-weapon States

1	Ireland	1 Jul. 1968
2	Nigeria	27 Sep. 1968
3	Denmark	3 Jan. 1969
4	Canada	8 Jan. 1969
5	United Rep. of Cameroon	8 Jan. 1969
6	Mexico	21 Jan. 1969
7	Finland	5 Feb. 1969
8	Norway	5 Feb. 1969
9	Ecuador	7 Mar. 1969
10	Mauritius	25 Apr. 1969
11	Botswana*	28 Apr. 1969
12	Mongolia	14 May 1969
13	Hungary	27 May 1969
14	Poland	12 Jun. 1969
15	Austria	27 Jun. 1969
16	Iceland	18 Jul. 1969
17	CSSR	22 Jul. 1969
18	Bulgaria	5 Sep. 1969
19	New Zealand	10 Sep. 1969
20	Syrian Arab Rep.	24 Sep. 1969
21	Iraq	29 Oct. 1969
22	German Dem. Rep.	31 Oct. 1969
23	Swaziland*	11 Dec. 1969
24	Nepal*	5 Jan. 1970
25	Sweden	9 Jan. 1970
26	Taiwan, China*	27 Jan. 1970
27	Iran	2 Feb. 1970
28	Afghanistan	4 Feb. 1979
29	Romania	4 Feb. 1970
30	Paraguay	4 Feb. 1970
31	Ethiopia	5 Feb. 1970
32	Malta*	6 Feb. 1970
33	Cyprus	10 Feb. 1970
34	Mali, Rep. of	10 Feb. 1970
35	Jordan	11 Feb. 1970
36	Lao People's Dem. Rep.*	20 Feb. 1970
37	Togo*	26 Feb. 1970
38	Tunisia	26 Feb. 1970
39	Yugoslavia	3 Mar. 1970
40	Burkina Faso*	3 Mar. 1970
41	Costa Rica	3 Mar. 1970
42	Peru	3 Mar. 1970
43	Malaysia	5 Mar. 1970
44	Jamaica	5 Mar. 1970
45	Liberia	5 Mar. 1970
46	Somalia*	5 Mar. 1970
47	Greece	11 Mar. 1970
48	Maldives*	7 Apr. 1970
49	Ghana	5 May 1970
50	Lesotho*	20 May 1970

APPENDIX C (continued)

Non-nuclear-weapon States

51	Bolivia	26 May 1970	
52	Haiti	2 Jun. 1970	
53	Kenya	11 Jun. 1970	
54	Lebanon	15 Jul. 1970	
55	Zaire	4 Aug. 1970	
56	San Marino*	10 Aug. 1970	
57	Uruguay	31 Aug. 1970	
58	Guatemala	22 Sep. 1970	
59	Madagascar	8 Oct. 1970	
60	Central African Rep.*	25 Oct. 1970	acc
61	Morocco	27 Nov. 1970	
62	Senegal	17 Dec. 1970	
63	Holy See	25 Feb. 1971	acc
64	Chad*	10 Mar. 1979	
65	Burundi*	19 Mar. 1971	acc
66	Tonga*	7 Jul. 1971	
67	Dominican Rep.	24 Jul. 1971	
68	Dem. Kampuchea	2 Jun. 1972	acc
69	El Salvador	11 Jul. 1972	
70	Fiji*	14 Jul. 1972	acc
71	Philippines	5 Oct. 1972	
72	Benin*	31 Oct. 1972	
73	Thailand	7 Dec. 1972	acc
74	Australia	23 Jan. 1973	
75	Nicaragua	6 Mar. 1973	
76	Côte d'Ivoire	6 Mar. 1973	
77	Honduras*	16 May 1973	
78	Bahamas*	10 Jul. 1973	acc
79	Sudan	31 Oct. 1973	
80	Gabon	19 Feb. 1974	acc
81	Grenada*	19 Aug. 1974	acc
82	Sierra Leone	26 Feb. 1975	acc
83	Western Samoa*	17 Mar. 1975	acc
84	Korea, Rep. of	23 Apr. 1975	
85	Belgium	2 May 1975	
86	Germany, Fed. Rep. of	2 May 1975	
87	Italy	2 May 1975	
88	Luxembourg	2 May 1975	
89	Netherlands	2 May 1975	
90	Gambia*	12 May 1975	
91	Rwanda*	20 May 1975	acc
92	Libyan Arab Jama.	26 May 1975	
93	Venezuela	26 Sep. 1975	
94	Singapore	10 Mar. 1976	
95	Japan	8 Jun. 1976	
96	Suriname*	30 Jun. 1976	succ
97	Guinea Bissau*	20 Aug. 1976	acc
98	Panama	13 Jan. 1977	
99	Switzerland	9 Mar. 1977	
100	Portugal	15 Dec. 1977	acc
101	Liechtenstein	20 Apr. 1978	acc
102	Congo, People's Rep.*	23 Oct. 1978	acc
103	Tuvalu*	19 Jan. 1979	succ

APPENDIX C (continued)

Non-nuclear-weapon States			
104	Sri Lanka	5 Mar. 1979	
105	Democratic Yemen*	1 Jun. 1970	
106	Indonesia	12 Jul. 1979	
107	Bangladesh	27 Oct. 1979	acc
108	Cape Verde*	24 Oct. 1979	acc
109	St. Lucia*	28 Dec. 1979	acc
110	Barbados*	21 Feb. 1980	
111	Turkey	17 Apr. 1980	
112	Egypt	26 Feb. 1981	
113	Solomon Islands*	17 Jun. 1981	succ
114	Antigua & Barbuda*	1 Nov. 1981	succ
115	Papua New Guinea*	25 Jan. 1982	acc
116	Nauru*	7 Jun. 1982	acc
117	Viet Nam, Soc. Rep.	14 Jun. 1982	acc
118	Uganda	20 Oct. 1982	acc
119	Dominica*	10 Aug. 1984	succ
120	Equatorial Guinea*	1 Nov. 1984	acc
121	St. Vincent & the Grenadines*	6 Nov. 1984	succ
122	Brunei Darussalam*	26 Mar. 1985	acc
123	Seychelles*	12 Mar. 1985	acc
124	Bhutan*	23 May 1985	acc
125	Kiribati*	18 Apr. 1985	succ
126	Belize*	9 Aug. 1985	succ
127	Korea, Dem. People's Rep.	12 Dec. 1985	acc
128	Malawi*	18 Feb. 1986	succ
129	Colombia	8 Apr. 1986	
130	Yemen Arab Rep.*	14 May 1986	
131	Trinidad and Tobago*	30 Oct. 1986	
Depository Governments			
1	UK	27 Nov. 1968	
2	USA	5 Mar. 1970	
3	USSR	5 Mar. 1970	

*Non-member of the IAEA

Source: International Atomic Energy Agency, Vienna, Austria, December 1986.

INDEX